FLORA OF TROPICAL EAST AFRICA

DENNSTAEDTIACEAE

BERNARD VERDCOURT

Erect or climbing terrestrial or epilithic ferns, often large, with creeping or erect rhizomes covered with hairs or hair-like non-peltate scales and with single or complex double solenosteles, or prostelic. Fronds 1–4-pinnate or 2–4(–5)-pinnatifid, sometimes with continuous apical growth (*Hypolepis*); stipes with undivided or dissected U-shaped vascular strands or 2 C-shaped strands back to back; lamina glabrous or hairy, veins free or anastomosing. Sori marginal, submarginal or superficial near the margin, ± round to oblong or linear (on one or both margins if marginal). Indusium absent or cup-shaped and opening outwards (*Microlepia*) or linear, or with leaf-margin modified to form a false indusium (*Hypolepis*), or both true and false indusia present; paraphyses present or absent. Spores monolete or trilete.

As defined here (following Kramer in Kubitzki et al., Fam. Gen. Vasc. Pl.: 81–94 (1990) with slight changes) the family includes 17 genera, 9 of which occur in East Africa. There is, however, much disagreement over the delimitation of the group. *Lindsaeaceae* Pic. Serm. is frequently kept separate but Kramer gives it as subfamily *Lindsaeoideae*, without giving a Latin description. Pichi-Sermolli (Webbia 24: 705 (1970)) also separates *Hypolepidaceae* which covers genera 2, 3, 4, 5 and 6 in this account.

1. Climbing ferns; sori terminal on single veins, immersed in
    tissue (**U 2**) ·   7. **Odontosoria**
   Erect ferns · 2
2. Veins anastomosing; fronds not simply pinnate, or if so,
    robust, broad and usually hairy · 3
   Veins mostly free or, if anastomosing somewhat (in *Lindsaea
    ensifolia*), then fronds simply pinnate, slender, narrow and
    glabrous · 4
3. Fronds pubescent to hirsute at least on the costae and axes · ·   4. **Blotiella**
   Fronds glabrous, and glaucous when young · · · · · · · · · · · ·   5. **Histiopteris**
4. Sori not marginal, subcircular, opening towards the margin,
    frequently corresponding with a hydathode on the upper
    surface ·   1. **Microlepia**
   Sori marginal or apparently marginal · 5
5. Sori with outer false indusium formed by reflexed modified
    leaf-margin and sometimes with inner true indusium · · · · · · · · · · · · · · 6
   Sori with only a true indusium · 8
6. Sori short on single veins, protected by a short rounded
    marginal reflexed flap ·   3. **Hypolepis**
   Sori longer, covering several to many veins, the outer false
    indusium occupying most or ± all of the margins of the
    fertile segments · 7

1

7. Stipe with 2 vascular bundles; fronds slightly succulent but drying ± membranous; paraphyses present; rachis without callose spots (nectaries) ·   **6. Lonchitis**
   Stipe with several vascular bundles; fronds coriaceous; paraphyses absent; rachis with callose spots (nectaries) at insertion of larger pinnae ·   **2. Pteridium**
8. Ultimate pinnules almost linear, up to 4 mm long, uninerved, each with only one sorus; indusium pocket-like, attached at base and sides; rhizome short-creeping · · · · · · · · · · · · ·   **8. Sphenomeris**
   Ultimate pinnules either cuneate, 4–13 mm long with sori on 1–4 vein-ends interrupted by marginal incisions (*L. madagascariensis*) or linear to lanceolate or oblong, cuneate at base, (3.5–)10–22 cm long in pinnate fronds or lamina simple, linear or lanceolate up to 10 cm long with long linear continuous sori on both margins (*L. ensifolia*); indusium short to elongate, opening outwards; rhizome long-creeping ·   **9. Lindsaea**

1. MICROLEPIA

C. Presl, Tent. Pterid.: 124 (1836) & in Abh. Königl. Böhm. Ges. Wiss., ser. 4, 5: 124 (1837); Kramer in Kubitzki et al., Fam. Gen. Vasc. Pl. 1: 84 (1990)

Rhizome creeping, covered with brown multicellular hairs. Fronds often quite large, mostly closely spaced; stipe with a single U-shaped bundle; lamina 1–3-pinnate to 4(–5)-pinnatifid, venation free. Sori intramarginal, ± round, borne on a vein ending and with a small membranous cup- or pouch-shaped indusium opening outwards and attached at its base and sides. Spores trilete, tetrahedral-globose, echinate.

About 45 species mainly in the Old World tropics, including one common pantropical species.

Ultimate segments equally lobulate on both margins; fronds thinner and drying pale yellowish green (widespread) · · ·   **1. *M. speluncae***
Ultimate segments straight or less lobulate on basiscopic side; fronds thicker and drying darker green (**T** 6, 7) · · · · · · · ·   **2. *M. fadenii***

1. **M. speluncae** (*L.*) *Moore*, Ind. Fil. 1: 93 (1857); Kuhn, Fil. Afr.: 159 (1868); Hieron. in P.O.A. C: 77 (1895); V.E. 2: 21, fig. 18 (1908); Sim, Ferns S. Afr., ed. 2: 129, t. 38 (1915); F.D.-O.A.: 66 (1929); Tardieu in Mém. I.F.A.N. 28: 58, t. 7/3–5 (1953) & Fl. Madag. 5(1): 9 (1958); Alston, Ferns W.T.A.: 33 (1959); Tardieu, Fl. Cameroun 3, Ptérid.: 94, tt. 11/3–5, 15/3–5 (1964) & Fl. Gabon 8, Ptérid.: 68, t. 11/3–5 (1964); Schelpe, F.Z., Pterid.: 89, t. 27 (1970) & Expl. Hydrobiol. Bassin L. Bangweolo & Luapula, 8(3), Ptérid.: 49, fig. 17 (1973) & C.F.A., Pterid.: 75, t. 13 (1977); Schelpe & Diniz, Fl. Moçamb., Pterid.: 91, t. 8 (1979); W. Jacobsen, Ferns S. Afr.: 210, fig. 147, map 45 (1983); Pic. Serm. in B.J.B.B. 53: 258 (1983); Schelpe & N.C. Anthony, F.S.A., Pterid.: 85, fig. 24*, map 68 (1986); Benl, Pterid. Bioko 4, in Acta Bot. Barcinon. 38: 23 (1988); J.E. Burrows, S. Afr. Ferns: 106, t. 16/4, fig. 23/106, map (1990); Faden in U.K.W.F., ed. 2: 24 (1994). Type: Sri Lanka, *Herman* (BM-HERM, vol. 3, fol. 41!, lecto.)

Terrestrial fern; rhizome much branched, ± 1 cm diameter, multicellular hairs pale, up to 4 mm long; fronds spaced up to 6 cm apart, (0.45–)1.5–3 m tall; stipe 0.2–1 m long and 6 mm wide, at first pubescent, later glabrous; lamina thin, yellow-

* Captions to figs. 23 and 24 are transposed.

FIG. 1. *MICROLEPIA SPELUNCAE* — **1**, pinna, × ²/₃; **2**, fertile pinnule segment, × 2. Both from *Fisher & Schweickerdt* 449. Drawn by Monika Shaffer-Fehre. From F.Z.

green, ovate, 0.4–1.8 m long, 0.1–1.4 m wide, 3–4(–5)-pinnatifid; pinnae alternate, narrowly oblong to oblong-lanceolate, up to 60 cm long and 21 cm wide, 2–3-pinnate to 2–3-pinnatifid, acute; pinnules alternate, up to 13 cm long and 6 cm wide; ultimate segments oblong, 4–12 mm long, 3–4 mm wide, crenate to pinnatifid, obtuse, thinly to fairly densely pubescent above and below with pale multicellular hairs up to 0.4 mm long; rachis thinly pubescent with minute hairs, eventually glabrous. Sori 2–10(–20) per segment, clearly intramarginal, round, 1 mm diameter, situated beneath on a vein ending which shows as a usually conspicuous hydathode on the upper surface; indusium often caducous, often pubescent-hispid. Fig. 1.

UGANDA. Toro District: Bwamba, Kabango, 21 Nov. 1935, *A.S. Thomas* 1501!; Masaka District: Sese Is., Towa Forest, 30 June 1935, *A.S. Thomas* 1355!; Mengo District: 16 km on Entebbe road (from Kampala), June 1937, *Chandler* 1681!
KENYA. North Kavirondo District: Malava [Kabras] Forest, W side of Kakamega–Broderick Falls road, 26 Nov. 1969, *Faden* 69/2036! & Kakamega Forest, along Yala R. about 5 km SE of Forest Station, 25 Nov. 1969, *Faden et al.* 69/2011!; Masai District: foot of Ngulia Hills, near stream pool, 28 Aug. 1969, *Bally* 13470!
TANZANIA. Lushoto District: Amani, Mt Bomole, 14 Oct. 1928, *Greenway* 889!; Ufipa District: Sumbawanga, Chapota, 7 Mar. 1957, *Richards* 8534!; Iringa District: Mwanihana Forest Reserve, above Sanje village, 10 Oct. 1984, *D.W. Thomas* 3932!
DISTR. U 2, 4; K ?3, 5–7; T 3, 4, 6, 7; throughout tropical Africa to S & SW Africa and widespread throughout the tropics
HAB. Forest edges, swamp forest, *Khaya-Cynometra* forest, ditch-sides, streams, old rubber plantations; (350–)700–1650(–1850) m

SYN. *Polypodium speluncae* L., Sp. Pl., ed.1: 1093 (1753)
    Davallia speluncae (L.) Baker, in Hook. & Baker, Syn. Fil.: 100 (1867)

NOTE. When growing in rocky places, crevices etc. it can be much smaller e.g. *Verdcourt* 57 & 58 (Lushoto District, R. Dodwe by Amani to Monga road crossing, 23 Jan. 1950). Several varieties have been described from Asia including var. *villosissima* C. Chr. with varying degrees of indumentum but I have considered the East African material typical and am not sure of the validity of these varieties. The lectotype is rather sparsely pilose. Material I have seen from K 3 has been wrongly identified, and was not this species. The lowest altitude and K7 record are based on a sterile young specimen (Kwale District: Shimba Hills, Mwele-Mdogo, 12 Sept. 1997, *Luke et al.* 4724!)

2. **M. fadenii** *Pic. Serm.* in Webbia 27: 406, fig. 6 (1973). Type: Tanzania, Uluguru Mts, Mwere Valley, 26 Sept. 1970, *Faden et al.* 70/619 (Herb. Pic. Serm., holo., EA, K!, iso.)

Terrestrial fern; rhizome tough, long-creeping, dichotomously branched, 7–11 mm diameter, with subulate multicellular thick rigid hairs up to 2.5 mm long. Fronds spaced at about 1.3 cm intervals, 1.6–2.5 m tall; stipe ± terete, sulcate above, ± 1 m long, 1 cm in diameter at base, puberulous; lamina ovate in outline, gradually narrowed to the apex, ± 1.5 m long, 80 cm wide, tripinnatisect; rachis puberulous above but at length glabrous, densely adpressed pilose at base; lower and middle pinnae narrowly lanceolate-triangular, 40 cm long, 7 cm wide, long-attenuate at apex; upper pinnae linear-lanceolate; rachis of pinnae entirely glabrous beneath; pinnules narrowly triangular or lanceolate-triangular, acute or long-attenuate at the apex; acroscopic segments subdimidiate, unequally obovate or obovate-rhombic, narrowed at base, rounded at apex, basiscopic side very reduced and 1-veined, margin usually entire, acroscopic side several-veined with curved 2–4-lobulate margin; lowest segments up to 10 mm long, 5 mm wide; veins evident on both surfaces, not reaching the margin, terminated by narrowly clavate hydathodes. Sori 1–3 in each segment, borne on acroscopic side of segments only, terminal on the veinlets, ± remote from margins; indusium densely bristly pubescent with hairs similar to those on underside of segments.

TANZANIA. Morogoro District: Uluguru Mts, Mwere Valley, 26 Sept. 1970, *Faden et al.* 70/619! & S Uluguru Mts, E slopes above Sumbini [Simbini] village, 14 Mar. 1971, *Pócs* 6418/D!; Iringa

District: Kilombero, Mwanihana Forest Reserve, above Sanje village, 10 Oct. 1984, *D.W. Thomas* 3869!

DISTR. **T** 6, 7; not known elsewhere

HAB. Evergreen rain-forest; 1400–1700 m

NOTE. This is related to *M. rhomboidea* (Hook.) Prantl, an Indian species, and quite distinct from *M. speluncae.*

2. **PTERIDIUM**

Scop., Fl. Carniol.: 169 (1760); R.M. Tryon in Rhodora 43: 1–31, 37–67 (1941); Kramer in Kubitzki et al., Fam. Gen. Vasc. Pl. 1: 85 fig. 34/A (1990), *nom. conserv.*

Erect fern; rhizome subterranean, extensively creeping and repeatedly branched, densely covered with multicellular chestnut hairs but not scaly. Fronds alternate, often large; stipe ± woody, elongate with numerous vascular bundles, felty just below ground but glabrous above; rachis with rather obscure callose spots at insertion of larger pinnae. Lamina 3–4-pinnate or 4-pinnatifid, eventually coriaceous, glabrous to densely pubescent or tomentose beneath; segments very numerous, linear to ovate, with revolute margins; veins free. Sori linear, submarginal on ultimate segments, mostly continuous, with an outer false indusium consisting of the reflexed margin and a delicate inner indusium which is continuous or interrupted but may be reduced to a few hairs; paraphyses absent; spores tetrahedral-globose trilete, brown, very finely spinulose.

Usually accepted to be a single species throughout the temperate and tropical regions but divided into numerous subspecies and varieties; some authors still accept more than one species. Bracken is very important ecologically; it can be a very bad weed and is poisonous. For earlier references the following is useful – K.W. Braid, Bracken: a review of the literature. C.A.B. Publication No. 3 (1959). Volume 73 (1976) of Bot. J. Linn. Soc. is devoted to 'The Biology of Bracken'.

P. aquilinum (*L.*) *Kuhn* in von der Decken, Reisen Ost-Afr. 3 (3): 11 (1879); Hieron. in P.O.A. C: 78 (1895); V.E. 2: 47 (1908); Sim, Ferns S. Afr., ed. 2: 264, t. 134 (1915); F.D.-O.A.: 48 (1929); R.M. Tryon in Rhodora 43: 12 (1941); U.O.P.Z.: 260 (1949); Tardieu in Mém. I.F.A.N. 28: 67 (1953); Alston in Estudos Ensarios Doc. Junta Invest. Ci. Ultramar. 12(3): 14 (1954); Tardieu, Fl. Madag. 5(1): 66 (1958); Alston, Ferns W.T.A.: 33 (1959); P.G. Taylor, British Ferns & Mosses: 92, fig. 18 (1960); Tardieu, Fl. Cameroun 3, Ptérid: 96, t. 12/5–6 (1964) & Fl. Gabon 8, Ptérid: 71, t. 12/5–6 (1964); Ivens, E. Afr. Weeds: 200, fig. 100 (1967); Verdc. & Trump, Common Poisonous Pl. E. Afr.: 200 (1969); Schelpe, F.Z., Pterid.: 88 (1970) & Expl. Hydrobiol. Bassin L. Bangweolo & Luapula, 8(3), Ptérid.: 47 (1973) & C.F.A., Pterid.: 72 (1977); Kornaś, Distr. Ecol. Pterid. Zambia: 82, map 50B (1979); Schelpe & Diniz, Fl. Moçamb., Pterid.: 88 (1979); W. Jacobsen, Ferns S. Afr.: 208 (1983); Schelpe & N.C. Anthony, F.S.A., Pterid.: 83 (1986); J.E. Burrows, S. Afr. Ferns: 103 (1990); Faden in U.K.W.F., ed. 2: 24 (1994). Type: *Filix femina*, Fuchs, Hist.: 596, mispr. 569 (1542) (lecto.)*

Gregarious terrestrial fern with rhizome 2.5 cm diameter when fresh, ± 7 mm wide when dry. Fronds borne at intervals of 1 cm or more, 0.3–1.8(–3) m tall, up to 0.6 m wide, coiled spirally when young, with dense chestnut hairs 1–1.5 mm long; stipe 15–40(–90?) cm long; lamina triangular to oblong-ovate in outline; pinnae ovate-triangular, up to 40 cm long, 15 cm wide, acute, the basal pinnae from half to as long as the lamina; pinnule segments linear to oblong, bluntly short- to long-

---

* Chosen by R.M. Tryon in 1941; later Tardieu-Blot (1964) made the more sensible choice of Hort. Cliff. 473 no. 6 (BM), but Tryon's choice must stand.

caudate, the larger deeply pinnatifid into obtuse narrowly oblong lobes ± glabrous or thinly pubescent to densely pubescent or tomentose above; rachis and secondary rachises pale brown, eventually glabrous. False indusium membranous, ± 0.5 m wide, ciliate.

NOTE. Extensive synonymy is given by Schelpe (1970).

Basal pinnae about half the length of the lamina; pinnules terminating in simply pinnatifid segment, largest pinnule segments shortly caudate; frond 4-pinnatifid to 4-pinnate; lower surface pubescent to tomentose ·   subsp. *aquilinum*

Basal pinnae about as long as the lamina; pinnules simply long-caudate and largest pinnule segments long-caudate or linear, entire; frond 3-pinnate to 4-pinnatifid; lower surface subglabrous to pubescent   subsp. *centrali-africanum*

subsp. **aquilinum**; Schelpe, F.Z., Pterid.: 88 (1970) & Expl. Hydrobiol. Bassin L. Bangweolo & Luapula, 8(3), Ptérid.: 47 (1973) & C.F.A., Pterid.: 74 (1977); Schelpe & Diniz, Fl. Moçamb., Pterid.: 90 (1979); Pic. Serm. in B.J.B.B. 53: 260 (1983); W. Jacobsen, Ferns S. Afr.: 208, fig. 146/a–b, map 44 (1983); Schelpe & N.C. Anthony, F.S.A., Pterid.: 83, fig. 22, map 67 (1986); J.E. Burrows, S. Afr. Ferns: 104, t. 15/5, fig. 22/105a–b, map (1990)

Frond 4-pinnatifid to 4-pinnate; basal pinnae about half the length of the lamina; pinnules terminating in simply pinnatifid segment, largest pinnule segments shortly caudate; lower surface pubescent to tomentose.

UGANDA. Karamoja District: Mt Moroto, Jan. 1959, *J. Wilson* 645!; Ankole District: Mitoma, July 1939, *Purseglove* 864!; Masaka District: Sese Is., Towa Forest, 22 July 1939, *A.S. Thomas* 3027!

KENYA. Machakos District: Mbooni Hills, 8 Oct. 1947, *Bogdan* 1182!; Kisumu–Londiani District: Tinderet Forest Reserve, 15 June 1949, *Maas Geesteranus* 4988!; Kwale District: Shimba Hills National Reserve, near the Kivukoni entrance, 22 Nov. 1971, *Bally & Smith* 14340!

TANZANIA. Moshi District: 13 km S of Moshi, Arusha Chini, Feb. 1965, *Beesley* 84!; Lushoto District: Amani, Drachenberg Plantation, 16 Jan. 1950, *Verdcourt* 45!; Iringa District: Mufindi, 1 km above L. Ngowasi [Ngwazi] dam, 11 Aug. 1971, *Perdue & Kibuwa* 11024!; Zanzibar, about 25 km along Chwaka road, 7 Feb. 1929, *Greenway* 1364!

DISTR. U 1–4; K 3–7; T 1–4, 6, 7, 8 (intermediate); Z; P; W Africa to Ethiopia through tropical Africa to South Africa, Madagascar, Comoro Is.; throughout Europe

HAB. Grassland including alpine meadows, seasonally wet grassland with scattered trees, forest margins, rain-forest clearings, woodland, scrub, thicket on coral rag and sandy soil at the coast, often in rocky places; a bad weed in cultivations; near sea level–2850(–3000?) m

SYN. *Pteris aquilina* L., Sp. Pl., ed.1: 1075 (1753)
    P. capense Thunb., Prodr. Pl. Cap.: 172 (1800). Type: South Africa, Cape of Good Hope, *Thunberg* (UPS, holo.)
    P. lanuginosa Willd., Sp. Pl., ed. 4, 5: 403 (1810). Type: Réunion, *Bory* 53 (B-WILLD 20023!, syn.)
    P. aquilina L. var. *lanuginosa* (Willd.) Hook., Sp. Fil. 2: 196 (1858)
    Pteridium aquilinum (L.) Kuhn var. *lanuginosum* (Willd.) Kuhn in Engl., Hochgebirgsfl. Trop. Afr.: 94 (1892)
    P. aquilinum (L.) Kuhn subsp. *capense* (Thunb.) C. Chr., Ind. Fil.: 591 (1906); Chiov., Racc. Bot. Miss. Consol. Kenya: 148 (1935)
    P. aquilinum (L.) Kuhn subsp. *typicum* R.M. Tryon in Rhodora 43: 15, map 1 (1941), *nom: illegit.*

NOTE. *Agnew & Timberlake* 11129 (Kenya, L. Naivasha, Crescent I., 1 July 1976), a sterile specimen originally determined as *Microlepia speluncae*, is a form of *P. aquilinum*; the habitat "forming dense patches in the papyrus" is an unusual one for this species.

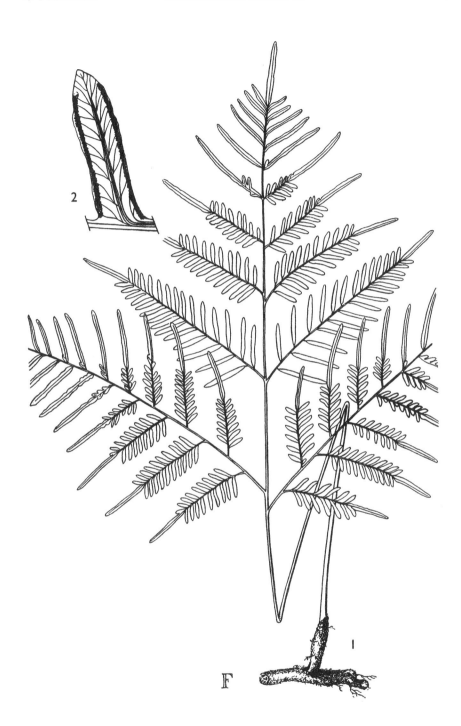

FIG. 2. *PTERIDIUM AQUILINUM SUBSP. CENTRALI-AFRICANUM* — **1**, habit, × ¹/₃; **2**, fertile pinnule segment, × 2. Both from *Drummond & Rutherford-Smith* 6935. Drawn by Monika Shaffer-Fehre. From F.Z.

subsp. **centrali-africanum** *Hieron.* in R.E. Fr., Wiss. Ergebn. Schwed. Rhod.–Kongo-Exped. 1: 7 (1914); Schelpe, F.Z., Pterid.: 89, t. 26 (1970) & Expl. Hydrobiol. Bassin L. Bangweolo & Luapula, 8(3), Ptérid.: 47, fig. 16 (1973) & C.F.A., Ptérid.: 74, t. 12 (1977); Kornaś, Distr. Ecol. Pterid. Zambia: 83, map 51b (1979); Schelpe & Diniz, Fl. Moçamb., Pterid.: 91, t. 7 (1979); W. Jacobsen, Ferns S. Afr.: 209, map 44 (1983); J.E. Burrows, S. Afr. Ferns: 104, t. 16/2, fig. 22/105 c–d, map (1990). Type: Congo (Kinshasa), Stanley Pool, *Hens* B59 (BM, lecto., not found)*

Frond 3-pinnate to 4-pinnatifid; basal pinnae about as long as the lamina; pinnules simply long-caudate and largest pinnule segments long-caudate or linear, entire; lower surface subglabrous to pubescent. Fig. 2.

Tanzania. Ngara District: Goya-goya plateau, 12 Nov. 1957, *Willan* 319!; Ufipa District: Mosi [Mozi] village, 2 Dec. 1958, *Richards* 10239!; Songea District: about 5 km E of Songea, by R. Luhira, 20 Mar. 1956, *Milne-Redhead & Taylor* 9253!
Distr. T 1, 4, 7, 8; Gabon, Congo (Kinshasa), Burundi, Angola, Zambia, Malawi, Mozambique and Zimbabwe
Hab. *Combretum* woodland, grassland, disturbed areas, a weed in cultivation; 1050–1800 m

Syn. *P. aquilinum* (L.) Kuhn subsp. *caudatum* (L.) Bonap. var. *africanum* Bonap., Not. Ptérid. 1: 62 (1915); R.M. Tryon in Rhodora 43: 51, map 6 (1941)**. Type: SW Tanzania, Ngaka ["Magaba-Thal"], *Busse* 944 (P, holo., EA!, iso., K!, fragment of iso.)
    P. centrali-africanum (Hieron.) Alston in Bol. Soc. Brot., sér. 2, 30: 22 (1956); Tardieu, Fl. Gabon 8, Ptérid.: 70, pl. 12/3–4 (1964); Pic. Serm. in B.J.B.B. 53: 261 (1983)

Note. Pichi-Sermolli considers this is undoubtedly a separate species but there do seem to be some intermediates. J. Thomson of NSW has been studying the species world-wide and his results will be published shortly.

3. HYPOLEPIS

Bernh. in Neues J. Bot. (Schrad.) 1(2): 34 (1806); Kramer in Kubitzki et al., Fam. Gen. Vasc. Pl. 1: 87 (1990)

Terrestrial plants or rarely on rocks, with widely creeping subterranean rhizomes with brown or reddish hairs. Fronds often large, 2–5-pinnatifid (or 2–4 pinnate and pinnatifid), hairy or glabrous and with free veins. Sori small, submarginal, usually protected by a reflexed marginal flap (pseudo-indusium) terminal on the veins, sometimes numerous enough to cover segment. Spores monolete.

A genus of 55 species, mostly pantropical or in the south temperate zone.

Stipe and rachis pale and smooth, pubescent to glabrous; lamina
    pubescent to glabrescent ·   1. *H. sparsisora*
Stipe and rachis purple-brown and rough with minute spinules;
    lamina with multicellular hairs · · · · · · · · · · · · · · · · · ·   2. *H. goetzei*

1. **H. sparsisora** (*Schrad.*) *Kuhn*, Fil. Afr.: 120 (1868); Sim, Ferns S. Afr., ed. 2: 236, t. 117 (1915); R.E. & T.C.E. Fr. in N.B.G.B. 9: 184 (1924); C. Chr. in Dansk Bot. Arkiv 7: 122 (1932); Chiov., Racc. Bot. Miss. Consol. Kenya: 145 (1935); Tardieu in Mém. I.F.A.N. 28: 59, t. 7/1–2 (1953) & Fl. Madag. 5(1): 8, fig. 1/1–2 (1958); Alston, Ferns W.T.A.: 33 (1959); Tardieu, Fl. Cameroun 3, Ptérid.: 95, t. 11/1–2 (1964); Schelpe, F.Z., Pterid.: 92, t. 28 (1970); Faden in U.K.W.F.: 30 (1974); Schelpe & Diniz, Fl. Moçamb., Pterid.: 94 (1979); W. Jacobsen, Ferns S. Afr.: 211, t. 148, map 46 (1983); Pic. Serm. in B.J.B.B. 53: 260 (1983); Schelpe & N.C. Anthony, F.S.A., Pterid.: 87, fig. 23***, map 69 (1986); J.E. Burrows, S. Afr. Ferns:

---

  * Tardieu-Blot cites *Hens* 59 as 'type', which has been accepted as lectotypification
 ** Schelpe (F.Z.) cites this as var. *africanum* "(Bonap.) Tryon"
*** Figs. 23 and 24 have been transposed so that captions do not apply.

106, t. 16/3, fig. 23/107, map (1990); Kramer in Kubitzki, Fam. Gen. Vasc. Pl. 1: 87, fig. 36 (1990); Faden in U.K.W.F., ed. 2: 24 (1994). Type: South Africa, Cape Province, *Hesse* (?LE, holo.)

Rhizome up to 5 mm in diameter, covered with multicellular brown hairs ± 1 mm long. Fronds erect, 20 cm or more apart with pinnae held horizontally, 1–1.8 m tall but occasionally forming thickets up to 3 m high; stipe straw-coloured, up to 1 m long, slightly or distinctly pubescent, usually finally glabrous; lamina 3–5-pinnatifid; pinnae up to 1 m long, usually less, ovate-deltate, with oblong acute crenate to pinnatifid adnate ultimate segments up to 1 cm long, glabrous except for a few scattered pale hairs on rachis branches and veins above and below, or slightly to distinctly grey-pubescent; rachis pale-brown, glabrous. Sori ± 1 mm in diameter, borne singly on the acroscopic margin of lobes of ultimate segments; false indusium semi-transparent, subentire. Fig. 3.

UGANDA. Toro District: Ruwenzori, Mobuku Valley, 7 Jan. 1939, *Loveridge* 328!; Kigezi District: below L. Mutanda, Nyamapana Swamp, 25 Mar. 1952, *Norman* 19! & L. Mutanda, Mushongero [Mushengera], 24 May 1963, *Kertland*!
KENYA. Naivasha District: Aberdares, S Kinangop, 24 Nov. 1957, *Molesworth-Allen* 3640!; Kiambu District: Gatamayu Forest, 8 Mar. 1964, *Verdcourt* 3989!; Teita District: Taita Hills, Chawia Bluff, Feb. 1955, *H.D. van Someren* 855!
TANZANIA. Moshi District: Kilimanjaro, near Bismarck Hut, 8 July 1956, *Pichi Sermolli* 5156!; Lushoto District: Amani, Bomole, 23 Nov. 1906, *Braun* 1432!; Morogoro District: NE Uluguru Mts, Kinole, 28 Sept. 1970, *Faden et al.* 70/673!; Mbeya District: Poroto Mts, Ngozi Crater, 16 Oct. 1956, *Richards* 6534!
DISTR. **U** 2; **K** 3, 4, 7; **T** 2, 3, 6, 7; Sierra Leone, W Cameroon, Bioko, São Tomé, Congo (Kinshasa) (Kivu), Rwanda, Burundi, Ethiopia, Malawi, Mozambique, Zimbabwe, South Africa, Madagascar and Mascarenes
HAB. Wet upland and montane evergreen forest with *Podocarpus, Neoboutonia, Trichocladus, Lasianthus, Ocotea, Macaranga, Hagenia, Ilex* etc., streamside and swampside forest and thicket, sometimes in gullies and clearings; 900–2800 m

SYN. *Cheilanthes sparsisora* Schrad. in Gött. Gel. Anz. 1818: 918 (1818)
   C. aspera Kaulf. in Linnaea 6: 186 (1831). Type: South Africa, Cape Peninsula, Table Mt, *Ecklon* (LZ†, holo., L, iso., BM!, BOL, photo.)
   [*C. anthriscifolia* sensu Schltdl., Adumbr. Pl.: 52 (1832) quoad spec. *Mundt & Maire*, non Willd.]
   Hypolepis aspera (Kaulf.) C. Presl, Tent. Pterid.: 162 (1836) & in Abh. Königl. Böhm. Ges. Wiss., ser. 4, 5: 162 (1837)
   Cheilanthes commutata Kunze in Linnaea 10: 542 (1836). Based on *C. anthriscifolia* sensu Schltdl.* Type: South Africa, Roodemuur, between Langekloof & Plettenbergsbai, Drège (?LZ†, holo., K!, iso.)
   [*Hypolepis punctata* sensu Hieron. in E.J. 28: 345 (1900) quoad *Goetze* 287, non (Thunb.) Kuhn]
   [*H. tenuifolia* sensu Peter, F.D.-O.A.: 42 (1929), pro parte, non Bernh.]

NOTE. This is by no means such a glabrous plant as sometimes implied Eggeling speaks of it as 'whole plant grey hairy' but it is glabrescent and lacks the rough indumentum of the next species; the stipes are never purple-brown.

2. **H. goetzei** *Reimers* in N.B.G.B. 12: 189 (1934); Faden in U.K.W.F., ed. 2: 24 (1994)**. Type: Madagascar, Ankafina (near Fianarantsoa), *Hildebrandt* 4139 (B, syn., BM!, isosyn.); between R. Mangoro and R. Matitana, Tanala, *Kitching* (K!, syn.); Pic d'Ivohibe, *Humbert* 3324 (BM-Herb. Christensen!, syn.) & 'mountains of Central and East Africa' (no specimens cited)

* This is not an illegitimate name as Schelpe claims; Kunze is giving a new name to the Cape element and specifically excludes the type of *C. anthriscifolia*.
** Taken up from a manuscript name of Hieronymus.

LMR

FIG. 3. *HYPOLEPIS SPARSISORA* — **1**, habit, × ²/₃; **2**, fertile pinnule segment, × 4. Both from *Swynnerton* 820. Drawn by Lura Mason Ripley. From F.Z.

Rhizome as in *H. sparsisora*. Fronds well-spaced, 0.7–2.1 m tall; stipe dark purplish or red, 30–60 cm long, rough with minute spinules and with stiff brown multicellular hairs; lamina triangular, 2–3-pinnatifid; rachis similar to stipe; pinnae triangular-lanceolate, 15–25 cm long, stalked, texture thicker than in last species; pinnules of third order elements divided into lanceolate or oblong lobed segments 0.5–2 cm long, the lobes oblique, rounded; midribs etc. with multicellular hairs. Sori with oblong or triangular green pseudo-indusia or sometimes naked (not in Africa).

UGANDA. Toro District: Ruwenzori, 4 Oct. 1905, *Dawe* 569! & Bujuku Valley, near Bigo Camp, 2 Apr. 1948, *Hedberg* 646!; Kigezi District: E Virunga Mts, between Muhavura and Mgahinga, 14 Nov. 1954, *Stauffer* 783!

KENYA. Northern Frontier Province: Mt Nyiru, 13 Dec. 1972, *Cameron* 139!; Trans-Nzoia District: NE Elgon, Apr. 1959, *Tweedie* 1810!; Nyeri District: Mt Kenya, Naro Moru Track, 6 Sept. 1963, *Verdcourt* 3727!

TANZANIA. Moshi District: Kilimanjaro, between Bismarck and Peter's Huts, June 1926, *Peter* 41958! & E Kilimanjaro, 20 Feb. 1971, *Vesey-FitzGerald* 6995!; Morogoro District: Uluguru Mts, Lukwangule Plateau, Nov. 1898, *Goetze* 287

DISTR. **U** 2; **K** 1, 3, 4, ?5, 6; **T** 2, 6; Bioko, Congo (Kinshasa) (Kivu), Ethiopia, St. Helena, Madagascar, Mascarenes

HAB. Boggy places and streamsides in upland forest, *Hypericum* woodland, giant heath association etc., (1500–)2100–3050(–3550) m

SYN. [*Hypolepis tenuifolia* sensu Peter, F.D.-O.A.: 42 (1929), pro parte, *non* Bernh.]
    H. rugosula (Labill.) J. Sm. var. *africana* C. Chr. in Dansk Bot. Arkiv 7: 121 (1932); Alston, Ferns W.T.A.: 33 (1959); Pic. Serm. in B.J.B.B. 53: 258 (1983). Types as for *H. goetzei*
    [*H. villoso-viscida* sensu Tardieu, in Fl. Madag. 5(1): 6, fig. 1/3–5 (1958) pro parte, *non* (Thouars) Tardieu, sensu stricto]
    [*H. rugosula* sensu Faden in U.K.W.F.: 29, fig. on 31 (1974), *non* (Labill.) J. Sm.]

NOTE. The exact name for this species will not be certain until the genus has been revised as a whole. A revision of the New Zealand species (Brownsey & Chinnock in New Zealand J. Bot. 22: 43–80 (1984)) has demonstrated that *H. rugosula* (Labill.) J. Sm. (as *rugulosa*) is an Australian endemic and does not occur in New Zealand. At one time its range was thought to include New Zealand, New Caledonia and Chile. Tardieu-Blot has used the earlier name *H. villoso-viscida* based on *Polypodium villoso-viscida* Thouars (Type: Tristan d'Acunha, *du Petit-Thouars* (P, holo.)). I have examined two sheets from Tristan preserved at Kew and find they differ in indumentum and pinnule-segments from East African material. Tardieu-Blot sinks *H. rugosula* into *H. villoso-viscida* and clearly considers the latter is an earlier name for the whole complex. I am also not certain the Réunion material (with no pseudo-indusium) is the same. Reimers cites var. *africana* in synonymy when establishing *H. goetzei* so the types are the same for both; *Schlieben* 4897 (B, ВM!) and *Goetze* 287 (B) cannot be the types. Tardieu-Blot does not cite *H. rugosula* var. *africana* in her synonymy but includes two of the syntypes in her citation of specimens. I have deliberately refrained from selecting a lectotype.

4. BLOTIELLA

A.F. Tryon in Contr. Gray Herb. 191: 96 (1962); Kramer in Kubitzki et al., Fam. Gen. Vasc. Pl. 1: 88 (1990)

Rhizome often large, thick and woody with reddish brown or golden hairs, often very long and dense. Fronds tufted, often large; stipes sulcate above, with several vascular bundles in a U-shape; lamina 1–3-pinnate, usually pubescent or hairy at least on the upper surface; ultimate segments entire to sinuate or deeply lobed or crenate; veins anastomosing. Sori marginal, either confined to bases of sinuses or extending around them and lobes, sometimes extensively; true indusium absent but reflexed margin of segment forms a membranous false indusium; paraphyses present. Spores monolete.

A genus of about 15 species in tropical and southern Africa, Madagascar and the Mascarenes with one in America. The species are very difficult and probably too many have been described. The division of the fronds is very variable and since the plants are usually large it is often difficult to make out the architecture in poorly collected material lacking adequate notes. Only R. Faden has prepared material which clearly shows the structure. At least some hybrids seem certain and part of the difficulty may be due to this. There is no doubt that this account is very unsatisfactory but I have refrained from taking the logical step of sinking all but species 1 and 7–9 into one species and adding to the confusion at this stage until spore structure etc. has been examined not only for all the African taxa but for those from Madagascar and the Mascarene Is. as well. In the meantime geography plays the largest rôle in choosing a name for a population. Young plants with simple to 1-pinnate fronds with few pinnae can only be named by association with adults in the same locality.

NOTE TO THE KEY: The upper pinnatifid (but not pinnate) pinnae towards the apex of a bipinnate frond must not be confused with the much smaller ultimate pinnules of the lower pinnae.

1. Pinnules distinctly coriaceous and shiny, narrowly triangular, 2.5–6.5 × 2–4 cm, undivided and entire with strongly revolute margin hiding the continuous sorus (T 6, Uluguru Mts) ··   9. *B. coriacea*
   Pinnules not so coriaceous and without other characters combined · 2
2. Paraphyses ending in a long strong subulate rigid hyaline hair; fronds pinnate with pinnae ± uniformly lobed, up to about 25 lobes on each side, ± uniform; sori 8–18 per lobe together with one in sinus or often ± continuous along whole margin of pinna; underside of pinnae and margins ± densely bristly-pubescent; rhizome erect (U 4, T 1) · · · · · · · · · · · ·   8. *B. trichosora*
   Paraphyses ending in a cell little different from basal cells or inflated or curved but not in a long hair and without other characters combined (see fig. 4, p. 14) · 3
3. Venation of ultimate lobes of pinnules impressed or plane above but raised beneath (a character easily appreciated with practice but not visible in juvenile material); pinnules of lower pinnae not stalked but many or all narrowly joined by ± winged rachis, less often free; essentially an upland species (U 2, K 4, 7, T 2, 5, 6) ·   1. *B. glabra*
   Venation of ultimate lobes of pinnules narrowly raised on both surfaces; pinnules stalked or ± sessile but not joined (save as usual at apices of fronds/pinnae, and occasionally in *B. currorii* a lowland swamp forest species) · 4
4. Fronds essentially 1-pinnate; pinnae to 43 × 15 cm, pinnatifid, the lobes semicircular to lanceolate (occasionally one or two of the lowermost lobes ± free but not stalked); surface sparsely to distinctly adpressed pilose (K 7, Mbololo Hill; T 3, E Usambaras, Amani area) ·   4. *B. hieronymi*
   Fronds 2(–3) pinnate or if 1–pinnate then from Uganda or S and W Tanzania · 5
5. Fronds 2-pinnate; pinnules slightly stalked, the upper narrowly triangular-lanceolate, 4.5 × 1.5 cm, the lower ones more broadly triangular in outline, up to 7 × 4 cm, deeply pinnatifid with largest lobes lanceolate ± 2 cm long, crenulate; rachis and secondary axes rough to touch due to stiff hair-bases (U 4, Sese Is.) ·   7. *B.* sp. A
   Without such a conformation of pinnules; axes not so obviously rough or if so, pinnules uniformly lanceolate · · · · · · · · · · · · · · · · · · 6

6. Pinnules of bipinnate fronds usually clearly stalked (the stalks 2.5–10 mm long), 2–10(–17) × 1–6 cm, subentire to deeply crenate or lobed with about 7–10 lobes on each side of pinnule, the surface ± glabrous or with scattered hairs but not densely pilose; costae, axes etc. pubescent and often with ± adpressed blackish brown hairs (**K** 4, 7; **T** 3, E Usambaras, Lutindi; W Usambaras; **T** 6, Ulugurus, Ukagurus etc.) · · · ·   2. *B. stipitata*

    Pinnules of bipinnate fronds sessile or nearly so or narrowly joined by winged rachis; surface glabrous to ± densely pilose; or fronds 1-pinnate · 7

7.* Pinnule surface typically ± glabrous or much more sparsely pubescent; non-pinnate pinnae large, 15–35(–75) × 4–12 cm, deeply pinnatifid into acute lanceolate segments up to 2 cm wide (or often ± entire in W Africa); pinnules of 2-pinnate fronds 4.5–13 × 1.5–3.5 cm, undulate to crenulate or subentire; sori narrow, several (to ± continuous in W Africa etc.) under 1 mm wide; swamp forest (**U** 2, 4) · · · · · · · · · ·   3. *B. currorii*

    Pinnule surface thinly to quite densely hairy; non-pinnate pinnae 7.5–31 × 1.5–12 cm with segments mostly up to 1 cm wide; pinnules 3–7.5 × 1.5–2.2 cm, crenulate; sori round to reniform or lunate to U-shaped, over 1 mm wide; drier forest · · · · · · · · · 8

8. Stipe hairs with distinct thickened bases; pinnules 3–5.5 × 1.5–2 cm; non-pinnate pinnae 5–13 × 1.5–6 cm; always 2-pinnate (**U** 2, **T** 1) ·   6. *B. tisserantii*

    Stipe hairs without such distinct thickened bases; pinnules 4.5–7.5 × 1.5–2.2 cm; non-pinnate pinnae 7.5–31 × 1.5–12 cm; 1–2-pinnate (**U** 2, 4; **T** 4, 7) ·   5. *B. natalensis*

1. **B. glabra** (*Bory*) A.F. *Tryon* in Contr. Gray Herb. 191: 99 (1962); Schelpe, F.Z., Pterid.: 82 (1970) & Expl. Hydrobiol. Bassin L. Bangweolo & Luapula, 8(3), Ptérid.: 43 (1973); Kornaś, Distr. Ecol. Pterid. Zambia: 86 (1979); Schelpe & Diniz, Fl. Moçamb., Pterid.: 86 (1979); W. Jacobsen, Ferns S. Afr.: 204, fig. 143, map 41 (1983); Pic. Serm. in B.J.B.B. 53: 266 (1983); Schelpe & N.C. Anthony, F.S.A., Pterid.: 81, fig. 21/2, map 65 (1986); Benl, Pterid. Bioko 4, in Acta Bot. Barcinon. 38: 31 (1988); J.E. Burrows, S. Afr. Ferns: 102, t. 15/4, fig. 21/103, map (1990); Faden in U.K.W.F., ed. 2: 25 (1994). Type: Réunion, *Bory* 63 (P, holo., B-WILLD 20131, iso.)

Rhizome erect, ascending or creeping, thick and woody, up to 2.5 cm wide, branched, with a dense felt of golden or chestnut-brown hair up to 4 cm long. Fronds tufted or closely spaced (0.9–)2–3 m tall; stipe pale brown, said to be green with purple stripes when fresh, up to 80 cm tall with silky chestnut hairs at base, thinly pubescent or glabrous above; stipe and rachis often rugulose with persistent hair bases; lamina ovate-lanceolate to elliptic in outline, 1.3(–2) m long, up to 80 cm wide, 2–3-pinnate-pinnatifid, ultimately ± coriaceous; pinnae oblong-ovate to lanceolate-acuminate, 40–45 cm long, 13 cm wide, sessile, lowest pair ± reduced; costa with dense brownish hairs beneath; pinnules lanceolate-oblong, 10 cm long, 2.5 cm wide, usually at least upper 6–8 pairs connected by a narrow wing 2–3 mm wide on either side of costa of pinna; pinnules with ± 10–12 pairs of lobes, these ultimate segments broadly obtuse or rounded, 1.5 cm long, 0.8 cm wide, separated by broad sinuses, entire or lobulate, pubescent on both surfaces; venation plane or impressed above, conspicuously prominent beneath; rachis densely pubescent. Sori short, crescent-shaped or reniform, ± 4 mm long, at base of sinuses and also 1–3 smaller ones on

---

* If plant from Kenya (**K** 3, 4) try *B. stipitata* or its hybrid since the stipes are sometimes short or ± absent.

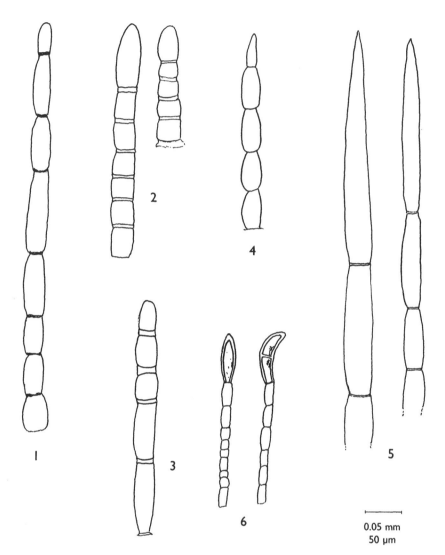

FIG. 4. *BLOTIELLA spp.* paraphyses — **1**, *B. GLABRA*; **2**, *B. NATALENSIS*; **3**, *B. TISSERANTII*; **4**, *B. SP. A*; **5**, *B. TRICHOSORA*; **6**, *B. CORIACEA*. Scale bar is 0.05 mm. 1 from *Pócs & Mabberley* 6739; 2 from *Hepper et al.* 5379; 3 from *Purseglove* 2259; 4 from *Thomas* 1349; 5 from *Chandler* 1745; 6 from *Pócs* 6477. Drawn by Bernard Verdcourt.

each margin above sinuses; sometimes the narrow costal wings have narrow sori ± 1.4 cm long; paraphyses sometimes apically recurved (according to literature) but those examined 0.4–0.6 mm long with long narrow cells at base and apical one short, straight, rounded or sausage-shaped.

UGANDA. Toro District: Ruwenzori, Mobuku Valley, 5 Jan. 1939, *Loveridge* 322! & Ruwenzori, 21 July 1960, *Livingstone, Kendall & Richardson* 14!; Kigezi District: Kishasha swamp, 26 Dec. 1961, *Morrison* 116!
KENYA. Fort Hall District: 6.5 km Thika–S Kinangop (Njabini) road, Kimakia Forest Station, 13 July 1969, *Faden & Evans* 69/896!; Meru District: NE Mt Kenya, Ithanguni, Kirui Cone, *Faden & Evans* 70/80A!; Teita District: Mt Kasigau, July 1937, *Dale* 3794!

TANZANIA. Moshi District: S Kilimanjaro, slopes between Umbwe and Weru Weru Rs, 30 Aug. 1932, *Greenway* 3179!; Kondoa District: near summit of Kinyassi Mt, 7 Feb. 1928, *B.D. Burtt* 1836!; Kilosa District: Ukaguru Mts, NNE slope of Mamiwa ridge, below Mnyera Peak, 30 July 1972, *Pócs & Mabberley* 6739c!

DISTR. **U** 2; **K** 4, 7; **T** 2, 5, 6; Cameroon, Equatorial Guinea, Bioko, São Tomé, Gabon, Congo (Kinshasa), Rwanda, Burundi, Ethiopia, Zambia, Mozambique, Zimbabwe, South Africa, Madagascar and Réunion (material so annotated from Mauritius is doubtful)

HAB. Mist forest and rain-forest of *Ocotea, Podocarpus, Cyathea* etc., also in giant heath zone with *Podocarpus* and *Hagenia*, sometimes in bamboo; 1350–3000 m

SYN. *Lonchitis glabra* Bory, Voy. Quatre Princ. Iles 1: 321 (1804); C. Chr. in Dansk Bot. Arkiv 7: 139, t. 54/1–2 (1932); Chiov., Racc. Bot. Miss. Consol. Kenya: 148 (1935); Tardieu, Fl. Madag. 5 (1): 78, fig. 11/6–9 (1958)

[*L. pubescens* sensu Oliv., in Trans. Linn. Soc. Bot. ser. 2, 2: 354 (1887) quoad *Johnston* 15, & sensu Sim, Ferns S. Afr., ed. 2: 261 pro parte quoad t. 132 (1915), *non* Kaulf.]

L. gracilis Alston in Exell, Cat. Vasc. Pl. S. Tomé, Suppl.: 7 (1956) & Ferns W.T.A.: 34 (1959); Tardieu, Fl. Gabon 8, Ptérid.: 78, t. 13/4–5 (1964) & Fl. Cameroun 3, Ptérid.: 102, t. 13/4–5 (1964). Type: Bioko [Fernando Po], near Moka, *C.D. Adams* 1100 (BM!, holo.)

[*Blotiella hieronymi* sensu Johns, Pterid. Trop. E. Afr.: 55 (1991) quoad *Livingstone* & *Kendall* 14, *non* (Kümmerle) Pic. Serm.]

2. **B. stipitata** (*Alston*) *Faden* in U.K.W.F.: 30, fig. 31 (1974); Faden in U.K.W.F., ed. 2: 25, t. 171 (1994). Type: Tanzania, Uluguru Mts, Morningside, *E.M. Bruce* 16 (BM!, holo.)*

Rhizome erect or shortly creeping, woody and much branched, covered with dense long or short dark brown golden or red-brown hairs up to 1 cm long. Fronds tufted or shortly spaced, (0.6–)1.5–2.7 m tall; stipe sometimes purplish, up to 1.5 m long; lamina simple in very young plants, becoming simply pinnate later, mature lamina subcoriaceous, up to 1.5 m long and 90 cm wide, variously divided, (1–)2(–3)-pinnate; uppermost segment of frond deeply pinnatifid; upper pinnae 9–14.5 cm long, 3.5–8 cm wide, unequally crenate-pinnatifid, 7–10 lobed on each side of costae, the lower lobes often larger, stalked, the stalks 7–10 mm long; lower pinnae pinnatifid at apex, pinnate at base, pinnules triangular-lanceolate, 2–17 cm long, 1–6 cm wide, narrowly acute at apex, stalked, the stalks usually 2.5–5 mm long (but subsessile in some **K** material); surfaces glabrous or with sparse hairs save for costae and costulae which are ± pubescent above and beneath; rachis and axes densely pale or dark brown spreading pubescent and typically (but not always) some adpressed or decumbent blackish hairs; occasionally all axes and surfaces ± glabrous. Sori elongate, crescent-shaped in deep sinuses and oblong to reniform, 3.5–4.5 mm long on subentire margins of pinnae/pinnule-lobes, up to 5 on each side; paraphyses ± 0.3 mm long with ± 8 thin-walled cells separated by thick walls, the apical cell ± 0.06 × 0.03 mm.

KENYA. Meru District: Nyambeni Hills, hill just S of Nyambeni Tea Estate, 12 Oct. 1960, *Verdcourt & Polhill* 2978! & Ithangune (Marimba) Forest, road along base of Kirui Cone, 13 km from Nkubu, 22 June 1969, *Faden et al.* 69/759!; Teita District: Voi, Sagala Hills, eastern slope, 1 Jan. 1971, *Faden, Evans & Kabuye* 71/39!

TANZANIA. Lushoto District: W Usambaras, Mazumbai forest reserve, 7 Jan. 1988, *Kisena* 592!; Kilosa District: Ukaguru Mts, 30 July 1972, *Pócs & Mabberley* 6742/A; Morogoro District: Morningside to Bondwa, 4 July 1970, *Faden et al.* 70/321!

DISTR. **K** 4, 7; **T** 3, 6; not known elsewhere

HAB. Wet evergreen forest including mist forest of *Parinari, Newtonia, Ocotea, Allanblackia, Myrianthus* etc. and of *Agauria, Drypetes, Ocotea, Chrysophyllum, Ilex mitis, Albizia, Macaranga* and *Syzygium; Podocarpus, Lasianthus, Xymalos* etc.; 1400–2400 m

SYN. *Lonchitis stipitata* Alston in Bol. Soc. Brot., sér. 2A, 30: 19 (1956)

* Most *Bruce* material is at Kew but it appears most of her ferns went to the BM.

NOTE. Some Kenya specimens, eg. *Faden* 74/1355, have been labelled subsp. nov. with a name that suggests 'much narrower', but this has not been published; this is also mentioned in a paper on Kilimanjaro ferns by A. Hemp (J. E. African Nat. Hist. Soc. 86: 38 (1999)) but without a description. I have not seen the Kilimanjaro material. The Kenya material certainly differs overall from that in the type locality.

B. stipitata × B. glabra

Pinnules not stalked; venation of segments not impressed above; rachis without black hairs.

KENYA. Meru District: NE Mt Kenya, Ithanguni, Kirui Cone, 28 Feb. 1970, *Faden & Evans* 70/80!; Teita District: Mt Kasigau, pipeline route from Rukanga, 6 Feb. 1971, *Faden et al.* 71/174!
DISTR. **K** 4, 7; not reported elsewhere (see note)
HAB. Montane forest; 1400–2100 m

SYN. *B.* sp. A; Johns, Pterid. Trop. E. Afr.: 55 (1991)
　　　B. sp. B; Johns, Pterid. Trop. E. Afr.: 55 (1991)
　　　B. sp. A; Faden in U.K.W.F., ed. 2: 25 (1994)

NOTE. There is some confusion about this hybrid. Faden says known only from Kirui Cone but states sp. A and B of Johns e.g. *Napper* 719, *Snowden* 592 (sphalm 719), *Faden et al.* 69/536 from Kimakia Forest, Limuru and locality unknown respectively, are this hybrid. Schelpe has determined material from N Malawi, Nyika plateau as this hybrid but it is not mentioned in F.Z. With the determination of species so highly unsatisfactory it is clear that the attribution of hybrids is often guesswork and virtually meaningless without clear evidence.

3. **B. currorii** (*Hook.*) *A.F. Tryon* in Contr. Gray Herb. 191: 99 (1962); Schelpe, F.Z., Pterid: 84 (1970) pro parte excl. specim. cit. & C.F.A., Pterid.: 68 (1977); Pic. Serm. in B.J.B.B. 53: 265 (1983); Benl, Pterid. Bioko 4, in Acta Bot. Barcinon. 38: 29 (1988); Kramer in Kubitzki, Fam. Gen. Vasc. Pl. 1: 88, fig. 37/a–b (1990). Type: Angola, near Elephants' Bay*, *Curror* (K!, syn.)

Rhizome erect or short-creeping, thick and woody, bearing dense ferruginous or golden hairs. Fronds tufted, 1.2–3(–3.6) m tall; stipe straw-coloured or brownish, up to 1 m with a dense felt of brown hairs up to 1 cm long at base, sparsely pubescent or ± glabrous above; lamina triangular, broadly ovate or oblong-lanceolate in outline, 0.5–2.5 m long, 45 cm wide, 1–2(–3?)-pinnate; pinnules ± sessile or joined by a narrowly winged rachis; rachis yellowish or brownish pubescent; terminal pinna segment broad, hastate and pinnatifid; pinnae usually large, 5–8 cm apart, broadly ovate in outline or broadly lanceolate, 15–35(–75) cm long, 4–6(–12) cm wide, acute or acuminate, ± entire (see note) to deeply pinnatifid, the segments subopposite, lanceolate, acuminate, 4–5 cm long, 1.5–2 cm wide, the lower $\frac{1}{2}$ or $\frac{2}{3}$ again sinuate or deeply lobed, pubescent, costae with similar indumentum to rachis; secondary pinnules of 2-pinnate fronds 4.5–13 cm long, 1.5–3.5 cm wide, entire, undulate or crenate. Sori narrow, continuous or interrupted, about 0.6 mm wide.

UGANDA. Masaka District: Sese Is., Bukasa I., Masekera, 26 Feb. 1933, *A.S. Thomas* 889!; Mengo District: Kipayu, Dec. 1914, *Dummer* 566! & edge of L. Nabugabo, Aug. 1935, *Chandler* 1307!
DISTR. **U** 2, 4; Mali, Sierra Leone, Liberia, Ivory Coast, Ghana, Togo, Benin, Nigeria, Cameroon, Equatorial Guinea, Bioko, Principe, São Tomé, Gabon, Central African Republic, Congo (Kinshasa), Burundi, Angola (see footnote)

* Hooker states that 'late Dr Curror' collected this at Elephants' Bay, western Africa about 32° S. This is clearly a confusion since no forest ferns could grow there. The locality appears to be another Elephants' Bay in Angola at 13°13' S. Dr A.B. Curror was on HMS Water-witch which stopped there in 1840 (see White & Sloane, The Stapelieae ed. 2, 3: 1059 (1937) and Mendonça in C.F.A. 1: xx (1937).

HAB. *Anthocleista, Macaranga, Neoboutonia* etc. swamp forest; 1100–1450 m

SYN. *Pteris* (*Litobrochia*) *currorii* Hook., Sp. Fil. 2: 232, t. 140/1–4 (1858), as '*currori*'; Baker in Hook. & Baker, Syn. Fil.: 168 (1868)
? *Pteris mannii* Baker in Hook. & Baker, Syn. Fil.: 168 (1868). Type: Bioko [Fernando Po], *Mann* (K!, syn.)
Lonchitis currorii (Hook.) Kuhn in von der Decken, Reisen Ost-Afr. 3 (3); 10 (1879); V.E. 2: 47, fig. 42 (1908); Tardieu, Mem. I.F.A.N. 28: 82 (1953) pro parte; Alston, Ferns W.T.A.: 34 (1959); Tardieu, Fl. Cameroun 3, Ptérid.: 100, fig. 13/1–3 (1964) & Fl. Gabon 8, Ptérid.: 74, fig. 13/1–3 (1964)
L. currorii var. *glabrata* Kümmerle* in Bot. Közlem. 14: 173 (1915). Type: Congo (Kinshasa), between Lusambo and Lomami, *Laurent* (BR, holo.)
? *L. mannii* (Baker) Alston in Bol. Soc. Brot., sér. 2, 30: 18 (1956) & Ferns W.T.A.: 34 (1959)
? *Blotiella mannii* (Baker) Pic. Serm. in Webbia 31: 250 (1977); Benl, Pterid. Bioko 4, in Acta Bot. Barcinon. 38: 30 (1988)

NOTE. Kornaś has pointed out that Schelpe's sole record of this species from the F.Z. area (*Mutimushi* 430) is referable to *B. crenata* (Alston) Schelpe. Ballard (K.B. 1937: 348 (1937)) considered *L. mannii* to be a juvenile 1-pinnate form of *L. currorii* and Schelpe also placed it in synonymy. Alston, however, treats it as a separate species and so does Benl based on the fact that *mannii* has a chromosome number 38 as against 76 for *currorii*. *B. mannii* with oblong undivided pinnae and sori continuous round the margin is distinctive but appears to me inseparable from *B. currorii*, there being every intermediate. No material has been seen from Uganda with continuous sori but much material in West Africa also has interrupted sori. I do not know on what material the record from Tanzania in C.F.A. is based.

4. **B. hieronymi** (*Kümmerle*) *Pic. Serm.* in Webbia 37: 131 (1983); Iversen in Symb. Bot. Upsal. 29 (3): 156 (1991). Type: Tanzania, E Usambaras, Gonja, *Holst* 4253 (W, holo., B, BM!, K!, P, iso.)

Rhizome erect or shortly creeping, base of stipe with dense felt of chestnut brown hairs at least 1 cm long. Fronds tufted, 0.6–2.5 m tall, simply pinnate or lowest pinnae with a few free pinnules; stipe up to 85 cm long, covered with short ± adpressed pale and dark hairs; lamina ± ovate-triangular, up to 80 cm long and wide; rachis with sparse to dense pale or brown hairs; pinnae in ± 10 pairs (1–4 in young plants), alternate or opposite, oblong-lanceolate to lanceolate, the upper about 10 cm long, 2 cm wide, ± sessile, the basal ones about 43 cm long, 15 cm wide, stalked, variously pinnatifid, acuminate at apex, the two apical pairs and terminal part of frond joined at their bases, somewhat bristly pilose-pubescent on venation which is pale and strongly raised beneath; stalks up to 1 cm long; pinnae with 20–30 pairs of lobes, those of upper pinnae rounded, 0.3–1.5 cm long, 0.7–1.2 cm wide, those of basal parts of lower pinnae much more elongate, triangular-lanceolate, up to 9 cm long, 1.8 cm wide, narrowly joined at base, narrowly acuminate at apex, slightly crenulate; lowest lobe sometimes elliptic and free; upper parts of lower pinnae with lobing similar to that of upper pinnae. Sori U-shaped in sinuses of small lobes, ± 5 mm long and often additional reniform ones above on one or both margins above sinus; false indusium very evident in young state; in large lobes with or without a sorus at the base of the sinus there can be up to 14 sori on each margin at crenulations; paraphyses short and numerous. Fig. 5.

KENYA. Teita District: Mbololo Hill, 5 July 1969, *Faden, Evans & Wolf* 69/844! & 12 Sept. 1970, *Faden, Evans & Smeenk* 70/563b! & 70/563c!
TANZANIA. Lushoto District: E Usambaras, Amani–Monga, 23 Jan. 1950, *Verdcourt* 63! & Bomole Hill, 30 June 1970, *Faden* 70/282! & Amani, 14 Sept. 1929, *Glynne* 218!
DISTR. K 7; T 3; not known elsewhere
HAB. Intermediate rain-forest e.g. *Allanblackia, Ocotea, Parinari* etc., mist forest of *Newtonia, Albizzia, Ocotea* etc.; 900–1450(–1600) m

---

* Christ's earlier use of this varietal epithet under *Pteris currorii* in Ann. Mus. Congo, Bot. sér. 2 (1): 67 (1899) is not accompanied by a description.

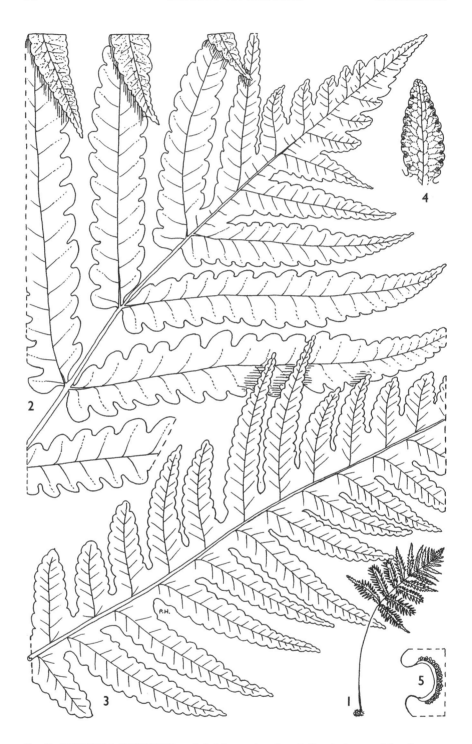

FIG. 5. *BLOTIELLA HIERONYMI* — **1**, habit, much reduced; **2**, apical part of frond, upper surface, × ²/₃; **3**, pinna, upper surface, × ²/₃; **4**, fertile pinnule, × ²/₃; **5**, sorus, much enlarged. All from *Faden* 70/282. Drawn by Pat Halliday.

SYN. *Lonchitis × hieronymi* Kümmerle in Bot. Közlem. 14: 174 (1915); Tardieu in Mém. I.F.A.N. 28: 84 (1953) pro parte
[*L. pubescens* sensu Peter, F.D.-O.A.: 48 (1929), *non* Kaulf.]

NOTE. Kümmerle considered this to be a hybrid between *Lonchitis currorii* (Hook.) Kuhn and *L. natalensis* Hook., but they do not grow in the E Usambaras; he mentioned the spores were abortive or few. He gives the type of his hybrid as *Holst* 4253 pro parte; on the same page (175) he refers *Holst* 4253 pro parte to *L. reducta* C. Chr. He also cites *L. pubescens* sensu Sim, Ferns S. Afr.: 75, t. 20 (1892) as a synonym of his supposed hybrid. The same plate in ed. 2 of Sim's work now plate 131 is cited by Schelpe & Anthony, F.S.A., Pterid.: 81 (1986) as *Blotiella natalensis* (Hook.) A.F. Tryon. One collection from Mbololo Hill had been determined as the latter by Schelpe before Faden redetermined it. Records in Johns, Pterid. Trop. E. Afr.: 55 (1991) of *B. hieronymi* from U 2, T 4 and K 7 are actually *B. glabra, B. natalensis* and *B. stipitata* respectively.

Tardieu-Blot (Mém. I.F.A.N. 28: 84, t.13/3–4 (1953)) records *Lonchitis hieronymi* from several places in West Africa citing 5 specimens but Alston (Ferns W.T.A.) makes no mention of this name nor does he cite any of the material cited by Tardieu-Blot, but she later (Fl. Cameroun 3, Ptérid.: 101 (1964)) mentions one of the cited specimens (*Annet* 213) under *L. currorii* but also makes no mention of her previous use of *L. hieronymi.*

5. **B. natalensis** (*Hook.*) *A.F. Tryon* in Contr. Gray Herb. 191: 99 (1962); Schelpe, F.Z., Pterid.: 82 (1970) & in Expl. Hydrobiol. Bassin L. Bangweolo & Luapula, 8(3), Ptérid.: 43 (1973) & in C.F.A., Pterid.: 68 (1977); Schelpe & Diniz, Fl. Moçamb., Pterid.: 87 (1979); Kornaś, Distr. Ecol. Pterid. Zambia: 86 (1979); W. Jacobsen, Ferns S. Afr.: 205, fig. 144, map 42 (1983); Schelpe & N.C. Anthony, F.S.A., Pterid.: 81, fig. 21/1, map 64 (1986); J.E. Burrows, S. Afr. Ferns: 100, t. 15.3, fig. 21/102, map (1990). Type: South Africa, Natal*, *Pappe* (K!, holo.)

Rhizome erect or creeping, often massive and woody, 1.5–2.5 cm diameter, covered with pale or reddish brown hairs 5–10 mm long. Fronds tufted or closely spaced, 0.6–2.6 m tall; stipe straw-coloured, up to 0.7(–1) m long with dense pale hairs when young but becoming glabrous; lamina lanceolate, narrowly oblong or triangular in outline, up to ± 1.5 m long, either 1-pinnate, 1-pinnate with a few lower pinnae having several free pinnules, or distinctly 2-pinnate; rachis pubescent, eventually slightly roughened by hair-bases, becoming glabrous below; terminal pinna ± hastate, pinnatifid and terminal segments of lateral pinnae similar with ± 7 pairs of lobes; pinnae narrowly oblong to narrowly triangular, (7.5–)18–31 cm long, (1.5–)3–12 cm wide, stalked, deeply pinnatifid with lobes lanceolate-triangular up to ± 4 cm long and 1 cm wide, narrowly joined at bases; in 2-pinnate fronds the pinnae have a deeply crenate terminal segment gradually passing into distinct pinnules which are narrowly triangular or lanceolate, 4.5–7.5 cm long, 1.5–2.2 cm wide, sessile, entire to wavy crenulate or pinnatifid; surfaces sparsely to densely pubescent with adpressed hairs, the costa with short pubescence even if surface glabrescent. Sori ± rounded, in sinuses between lobes of pinnules or pinnae but in more dissected fronds also in shallow sinuses of lobed margins of pinnules or pinna-lobes, up to ± 10 per pinnule and in places sometimes appearing almost continuous; paraphyses 0.2–0.4 mm long, thin, the apical cells brown, ellipsoidal, rounded or pointed.

UGANDA. Kigezi District: Ishasha Gorge, 6 km SW of Kirima, 21 Sept. 1969, *Faden et al.* 69/1210; Mengo District: 14 km from Kampala on Masaka road, May 1937, *Chandler* 1624! (see note on p.   )
TANZANIA. Buha/Kigoma District: Mkenke Valley, 26 Jan. 1964, *Pirozynski* 301!; Iringa District: Sao Hill, June 1958, *Watermeyer* 25!; Rungwe District: 15 km S of Tukuyu, Makete, 7 May 1975, *Hepper et al.* 5359!
DISTR. U 2, 4; T 4, 7; Congo (Kinshasa), Burundi, Angola, Zambia, Malawi, Mozambique, Zimbabwe, eastern South Africa, Madagascar, Comoro Is., Seychelles and Mauritius (*fide* Tardieu-Blot)

---

* Schelpe & Anthony state Durban but there is no evidence of this on the sheet itself. The type is at least bipinnate in part.

HAB. Streamside woodland and steep mountain forest with small streams, swampy area with
    Spondianthus, Uapaca and *Musanga,* secondary vegetation in regenerating forest cleared two
    to three years previously; 1100–1900 m

SYN. *Lonchitis natalensis* Hook., Sp. Fil. 2: 57, t. 89B (1858); Tardieu in Mém. I.F.A.N. 28: 82
    (1953) pro parte (excl. West African citations) & Fl. Madag. 5 (1): 72 (1958)
    [*L. pubescens* sensu Sim, Ferns S. Afr., ed. 2: 261, pro parte quoad t. 131 (1915), *non* Kaulf.*]

NOTE. I at first considered *B. crenata* Alston (Schelpe) to be a synonym of *B. natalensis* but
    Faden persuaded me this was wrong. Kornaś was convinced that *B. crenata* is a distinct species
    and pointed out that *Loveridge* 935 cited by Schelpe as *B. natalensis* was undoubtedly *B.
    crenata.* Specimens with 1-pinnate fronds, the pinnae oblong with about 20 rounded lobes on
    each side and ± densely hairy are admittedly distinctive and represent *B. crenata.* Distinctions
    such as rhizome erect *natalensis* and creeping *crenata* appear to be of little use. I do not know
    on what specimens Schelpe's record of *natalensis* from Kenya (in C.F.A.) is based but it must
    be admitted some Kenya material of *B. stipitata* can scarcely be distinguished.

6. **B. tisserantii** (*Alston & Tardieu*) *Pic. Serm.* in Webbia 37: 132 (1983); Benl,
Pterid. Bioko 4, in Acta Bot. Barcinon. 38: 33 (1988). Type: Central African
Republic, Oubangui, Boukoko, *staff of State Centre of Agric.* in *Tisserant* (P, holo.,
BM!, photo.)

Rhizome long and creeping, to 12 cm long, 1.5 cm diameter, densely brown-hairy
at apex. Fronds spaced, up to 1.5 m tall; stipe straw-coloured or brown, rather
slender, densely covered with spreading seta-like hairs 2.5 mm long with pustular
bases; lamina narrowly triangular or oblong, up to ± 1 m long, 40–50 cm wide, 2-
pinnate; non-pinnate pinnae 5–13 cm long, 1.5–6 cm wide; terminal pinna and
upper lateral pinnae of frond and also terminal parts of lower pinnae narrowly
triangular, pinnatifid; rachis slender, densely pubescent, slightly roughened by hair-
bases; non-pinnate pinnae 5–13 cm long, 1.5–6 cm wide; pinnules free, sometimes in
upper part of frond joined by a winged rachis, in about 7–11 pairs, oblong to oblong-
lanceolate, 3–5.5 cm long, 1.5–2 cm wide, rather shallowly crenate, the crenulations
mostly under 5 mm long, sessile or nearly so, ± densely covered with long white
decumbent hairs. Sori ± round to semilunate or U-shaped, small, in sinuses and
crenulations; paraphyses 0.2–0.3 mm long, thin, apical cell short and narrow.

UGANDA. Kigezi District: Ruhiza Forest, June 1956, *Milburn* 85! & Ishasha Gorge, Nov. 1946,
    Purseglove 2259! & Ishasha Gorge, Kitahulira, 17 June 1968, *Lock* 68/182!
TANZANIA. Bukoba District: Kaagya, *Gillman* 356!
DISTR. **U** 2; **T** 1; Nigeria, Bioko, Central African Republic; also recorded from Cameroon by
    Alston, and from Gabon by Tardieu-Blot
HAB. Forest; 1200–2250 m

SYN. *Lonchitis tisserantii* Alston & Tardieu in Mém. I.F.A.N. 28: 85, t. 13/5–6 (1953); Alston,
    Ferns W.T.A.: 34 (1959); Tardieu, Fl. Cameroun 3, Ptérid.: 102 (1964) & Fl. Gabon 8,
    Ptérid: 76 (1964)

NOTE. *Milburn* 85 had been determined by Alston as *L. sinuata* Alston (Type: Congo
    (Kinshasa), Eala, Munga, *Lebrun* 657 (BM!, holo.)) but this does not seem right and *L.
    sinuata* appears to me to be a variant of *B. currorii.* The material does on the other hand
    appear closely similar to the type of *B. tisserantii.* In her key Tardieu-Blot (p. 81) said the
    rhizome is long-creeping but in the description it is stated to be erect; the type appears to
    have a creeping rhizome, but without field notes it is difficult to be certain.

* References under this name in P.O.A. C, V.E. 2 etc. cannot be elucidated since no specimens
are cited.

7. **B.** sp. A

Rhizome not known. Fronds 1.8 m tall; stipe at least 30 cm long (probably twice as long), ± glabrous; lamina bipinnate with 11–12 pairs of pinnae (*fide* Thomas); pinnae narrowly triangular, ± 36 cm long, ± 14 cm wide with ± 10 free pinnules on either side; apical ones joined to form a pinnatifid terminal segment; rachis and secondary rachises pubescent with hairs which leave a stiff base when breaking off, the surface being slightly but distinctly rough to the touch; pinnules slightly stalked (stalk up to 3 mm long), upper ones narrowly triangular-lanceolate in outline, ± 4.5 cm long and 1.5 cm wide, crenately lobed, lobes up to ± 5 mm long; lower ones more broadly triangular in outline, up to 7 cm long and 4 cm wide, deeply pinnatifid with largest lobes lanceolate, ± 2 cm long, crenulate; surface with rather sparse appressed hairs, costa more densely hairy. Sori in sinuses of shorter lobes, ± U-shaped but up to 6 in the crenulations on each side of largest lobes and then lunate; paraphyses thin, 0.2–0.3 mm long, apical cell short, narrowly oblong and subacute.

UGANDA. Masaka District: Sese Is., Towa Forest, 30 June 1935, *A.S. Thomas* 1349! & Sese Is., Bugala I., 3 June 1932, *A.S. Thomas* 33!
DISTR. **U** 4; not known elsewhere
HAB. Primary rain forest; 1150–1200 m

NOTE. This had been identified as '*Lonchitis glabra*' but is most certainly not; I have been unable to ascribe it adequately to any species.

8. **B. trichosora** *Pic. Serm.* in B.J.B.B. 53: 266, fig. 15 (1983). Type: Burundi, Bururi, Kigwena, *Lewalle* 4403 (FT-Herb. Pic. Serm. (25305), holo., BR, iso.)

Rhizome erect, to 10 cm long, 1 cm diameter, covered with long chestnut hairs. Fronds tufted, 0.7–2 m tall; stipes straw-coloured, 30–75 cm long, with short and long hairs but eventually ± glabrous; lamina ovaté, ovate-elliptic or obovate in outline, pinnate or bipinnatipartite, 0.4–1.05 m long, 20–65 cm wide, pilose-pubescent above and ± densely bristly-pubescent beneath; pinnae in 10–16 pairs, lanceolate, 8–40 cm long according to position, distinctly stalked, acuminate, pinnatilobed, lobes up to about 25 pairs, 0.4–3 cm long (save lowest), ovate to oblong-triangular, rounded to ± acute at the apex, becoming smaller until apex of pinna is entire; basal lobes of pinna (in **U** specimen) twice as long as largest of rest and with stronger costae; lobes of pinnae crenulate. Sori oblong-reniform, in sinuses of pinna-lobes (± linear) and along margins of lobes in all but uppermost, the largest lobes with 8–18 in total excluding that in sinus, or ± continuous along whole pinna margin; paraphyses 0.5–0.6 mm long, articulate, terminated by a long rigid hair 0.2–0.3 mm long.

UGANDA. Mengo District: Kampala, Namanve Swamp, June 1937, *Chandler* 1745!; "Shores of the Victoria Nyanza", July 1878, *C.T. Wilson* 1!
TANZANIA. Bukoba District: near Bukoba, *Haarer* 2195! & just N. of Bukoba, around Kahororo secondary school, 11 Sept. 1974, *Balslev* 025!
DISTR. **U** 4; **T** 1; Burundi
HAB. Edge of swamp forest and swampy grassland on sandy lake deposits; 1100–1200 m

NOTE. A specialised species with very restricted distribution.

9. **B. coriacea** *Verdc.*, **sp. nov.** ab omnibus speciebus adhuc descriptis pinnulis distincte coriaceis, nitidis, anguste triangularibus, 2.5–6.5 × 2–4 cm, integris, margine revolutis; soris ambitu pinnularum margine continuis; cellulis apicalibus paraphysium brunneis crassimuralibus, ceteris hyalinis tenuimuralibus differt. Typus: Tanzania, S Uluguru Mts, above Kibungo mission, *Pócs & Lundqvist* 6477/c (EA!, holo., K!, photo.)

Rhizome and stipe not known. Lamina at least 70 cm long, bipinnate; rachis with dense dark brown curved hairs; pinnules narrowly triangular, 2.5–6.5 cm long, 2–4 cm wide, somewhat hastate at base, stalked (stalks 3–5 mm long), distinctly coriaceous with shiny surface, with brown hairs on midrib above and scattered dark brown curved hairs on surface beneath. Sorus apparently continuous, concealed by the strongly revolute margin; paraphyses with pale thin lower cells terminated by a deep brown straight or curved sausage-shaped part with thick walls of 1–2 cells ± 0.1 mm long.

TANZANIA. Morogoro District: S Uluguru Mts, eastern side ridge above Kibungo Mission, 19 Oct. 1971, *Pócs & Lundqvist* 6477/c!
DISTR. **T** 6; not known elsewhere
HAB. Montane heath; 1730 m

NOTE. This is a striking plant but the material is sparse and has a somewhat abnormal look, several of the pinnules being deformed.

Material still in doubt

Benl (Pterid. Bioko 4, in Acta Bot. Barcinon. 38: 32 (1988)) records *B. reducta* (C. Chr.) R.M. Tryon. Type: Guinea, Futa Djallon, Pita, *Pobéguin* 28 (BM-Herb. Christensen!, holo., P, iso., BM!, photo.)) from Uganda and Tanzania, but I have not seen the material concerned, although various sheets have been wrongly determined as this. The type and other young material are 1-pinnate and have the pinnae densely white-hairy on both surfaces; but much material with mostly 1-pinnate adult fronds annotated by Alston and others from much further away in Bioko, São Tomé etc. seems doubtfully separate from *B. currorii*. Johns (Pterid. Trop. E. Afr.: 55 (1991)) cites *Faden* 69/1037 (Uganda, E Mengo), 69/1208 (Uganda, Kigezi, Ishasha), 70/80 (Kenya, Mt Kenya) and *A.S. Thomas* 3675 (not found, but this number is listed as *Brachiaria*, a grass, in determination lists) under *B. reducta* stating that material from Uganda is possibly *B. tisserantii*.

There is also at Kew *Faden* 69/1209 and 69/1210 (same as 1208). Only 69/1208 is 1-pinnate and is a young plant with tufted fronds 30 cm tall and has been determined as a juvenile *B. natalensis* by Schelpe; all the rest of the Uganda material has also been determined as *B. natalensis* by him. Faden had suggested his 69/1208 and 69/1209 as *B. reducta* and 69/1208 had also been determined as ?*B. sinuata* (Alston) Pic. Serm. by Pichi Sermolli. I have at present accepted this material as *B. natalensis* sensu lato.

5. HISTIOPTERIS

(J. Agardh) J. Sm., Hist. Fil.: 294 (1875); Kramer in Kubitzki et al., Fam. Gen. Vasc.
Pl. 1: 89 (1990)

Erect fern; rhizome widely creeping, with brown hairs and/or scales. Fronds spaced; lamina 2–3-pinnatifid; pinnae opposite, sessile, with stipule-like basal pinnules or pinna-lobes, glabrous, glaucous, firmly herbaceous to thinly coriaceous; veins anastomosing. Sori marginal, continuous, with paraphyses; indusium formed from reflexed margin of lamina; spores monolete.

About 8 species, one pantropical, the others Asiatic.

H. incisa (*Thunb.*) *J. Sm.*, Hist. Fil.: 295 (1875); Hieron. in E.J. 28: 346 (1900); V.E. 2: 47, fig. 41 (1908); Sim, Ferns S. Afr., ed. 2: 263, t. 133 (1915); F.D.-O.A.: 48 (1929); Tardieu in Mém. I.F.A.N. 28: 67 (1953) & Fl. Madag. 5 (1): 68 (1958); Alston, Ferns W.T.A.: 34 (1959); Tardieu, Fl. Cameroun 3, Ptérid.: 98, t. 12/1–2 (1964) & Fl. Gabon

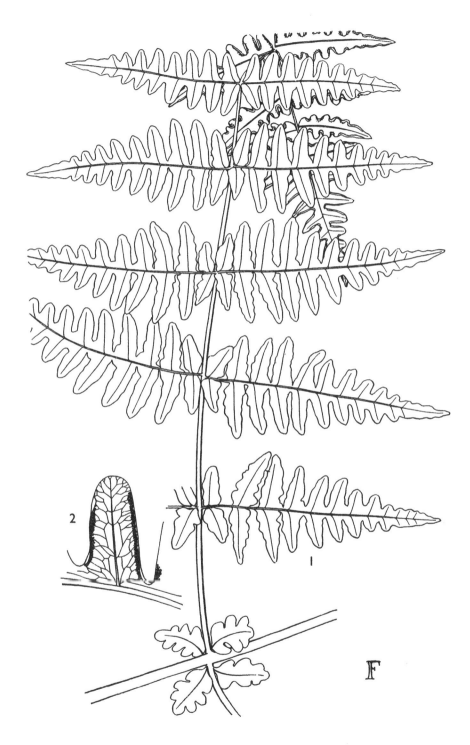

FIG. 6. *HISTIOPTERIS INCISA* — **1**, pinna, × ²/₃; **2**, fertile pinnule segment, × 2. Both from *Fisher & Schweikerdt* 336. Drawn by Monika Shaffer-Fehre. From F.Z.

8, Ptérid.: 72, t. 12/1–2 (1964); Schelpe, F.Z., Pterid.: 84, t. 24 (1970) & in Expl. Hydrobiol. Bassin L. Bangweolo & Luapula, 8(3), Ptérid.: 45, fig. 15 (1973); Faden in U.K.W.F.: 30 (1974); Schelpe, C.F.A., Pterid.: 70, t. 10 (1977); Kornaś, Distr. Ecol. Pterid. Zambia: 84, map 52a (1979); W. Jacobsen, Ferns S. Afr.: 206, fig. 145, map 43 (1983); Pic. Serm. in B.J.B.B. 53: 261 (1983) & 55: 201 (1985); Schelpe & N.C. Anthony, F.S.A., Pterid.: 82, fig. 21/3, map 66 (1986); J.E. Burrows, S. Afr. Ferns: 103, t. 16/1, fig. 23/104, map (1990); Kramer in Kubitzki et al., Fam. Gen. Vasc. Pl. 1: 89, fig 37/d–e (1990); Faden in U.K.W.F., ed. 2: 24 (1994). Type: South Africa, Cape Province, Grootvadersbosch, *Thunberg* (UPS, holo.)

Rhizome ± 5 mm in diameter, subterranean, with widely spaced fronds and a dense felt of brown multicellular hairs and/or scales (see Burrows p. 103). Fronds erect to arching with pinnae held horizontally, 0.6–2 m (occasionally 3 m) tall; stipe purplish brown or chestnut brown, up to 60 cm long, glabrous; lamina up to 40 cm long and 19 cm wide, pinnatifid to 2-pinnatifid, broadly lanceolate to ovate-triangular in outline, acute, and with basal pinna segments auriculate and developed very close to the rachis; ultimate lobes oblong, acute to obtuse, adnate, entire to sinuate, glaucous when young; rachis reddish-brown to pale brown nearer apex, terete, glabrous. Sori up to 1 mm broad at maturity, linear or rarely lunulate, borne along the margins of the ultimate lobes save at apices to lobes and sinuses; indusium membranous, entire. Fig. 6.

UGANDA. Toro District: Ruwenzori, Bujuku Valley, Aug. 1931, *Fishlock & Hancock* 138!; Kigezi District: Mt Mgahinga, 24 Aug. 1938, *A.S. Thomas* 2474!; Masaka District: Sese Is., Bubekke, 4 July 1935, *A.S. Thomas* 1380!

KENYA. South Nyeri District: Aberdare Mts, Kiandongoro Track above Tucha, at crossing of small tributary of R. Chania, 24 Oct. 1971, *R.B. & A.J. Faden* 71/882! & Aberdare Mts, *Ramsden*! & Mt Kenya, Castle Forest Station to Kamweti Track, 25 Jan. 1969, *Faden* 69/085!

TANZANIA. Bukoba District: Bukoba, Aug. 1931, *Haarer* 2196!; Moshi District: Kilimanjaro, S slope between Umbwe and Weru Weru Rs., 30 Aug. 1932, *Greenway* 3171!; Morogoro District: Uluguru Mts, Tanana, *Bruce* 76!; Iringa District: Sao Hill, June 1958, *Watermeyer* 27!

DISTR. U 2, 4; K 4; T 1, 2, 6, 7; very widespread in tropical and South Africa, pantropical and extending to S temperate and subantarctic islands

HAB. *Hypericum* woodland, *Podocarpus*-bamboo and giant heath-*Hagenia* forest and woodland, also swamp forest, rocky lake edges, ditch banks etc.; 1100–3450 m

SYN. *Pteris incisa* Thunb., Prodr. Fl. Cap.: 171 (1800); Baker in Hook. & Baker, Syn. Fil.: 172 (1868)

NOTE. Populations on Tristan da Cunha and Gough Is. have been referred to *H. incisa* var. *carmichaeliana* (J. Agardh) C. Chr.
Further synonymy is provided by Schelpe (1970).

6. LONCHITIS

L., Sp. Pl.: 1078 (1753) & Gen. Pl., ed. 5: 485 (1754); A.F. Tryon, Contr. Gray Herb. 191: 93–110 (1962); Lellinger in Taxon 26: 578–580 (1977); Kramer in Kubitzki et al., Fam. Gen. Vasc. Pl.: 89 (1990)

Anisosorus Maxon, Pterid. Porto Rico & Virgin Is.: 429 (1926), *nom. superfl.*

Rhizomes subterranean, creeping, covered with flattened several-celled brown hairs. Fronds rather closely spaced, slightly succulent but drying thin, 1–3-pinnate and pinnatifid; stipe with 2 C-shaped bundles uniting to form 1 bundle or sometimes additional bundles present; lamina with veins usually free or occasionally fused to form a few large areoles along the costa, with no included veins. Sori marginal, elongate, linear or curved in or around sinuses of frond segments; outer indusium membranous, formed by reflexed margin, opening inwards, inner indusium absent; paraphyses present; spores globose, trilete, granular.

FIG. 7. *LONCHITIS OCCIDENTALIS* — **1**, apical portion of frond, × ²/₃; **2**, lower portion of frond, × ²/₃; **3**, lower surface of fertile pinnules, × ²/₃. 1, 3 from *D.W. Thomas* 3683; 2 from *Hieronymus* 4233. Drawn by Pat Halliday.

A genus of 2 species, 1 African, the other American. Other species frequently referred to this genus are now placed in *Blotiella*.

L. occidentalis *Baker* in Hook. & Baker, Syn. Fil.: 128 (1867); Carruth., Cat. Afr. Pl. Welw. 2: 266 (1901); V.E. 2: 47 (1908); Tardieu, Fl. Madag. 5 (1): 79 (1958); Schelpe, F.Z., Pterid.: 86, t. 25 (1970) & Expl. Hydrobiol. Bassin L. Bangweolo & Luapula, 8(3), Ptérid.: 45 (1973) & C.F.A., Pterid.: 72, t. 11 (1977); Kornaś, Distr. Ecol. Pterid. Zambia: 85, map 52b (1979); Benl, Pterid. Bioko 4, in Acta Bot. Barcinon. 38: 34 (1988); Kramer in Kubitzki et al., Fam. Gen. Vasc. Pl.: 89, fig. 35/d–e (1990). Type: Angola, Cuanza Norte, Golungo Alto, Sobado de Quilombo, Quiacatubia, *Welwitsch* 132 (K! lecto., BM!, LISU, isolecto.)*

Rhizome short- to long-creeping, ± fleshy, 6 mm diameter when dry, covered with pale brown hairs. Fronds 1–2 cm apart, 0.3–1.6(–2) m tall; stipe straw-coloured to red or purple, thick, 12–50 cm long, covered with mauve and/or white hairs; lamina somewhat fleshy, 18–50 cm long, pinnate-pinnatifid to bipinnate-tripinnatifid, oblong to triangular-ovate, acute to acuminate, with reddish or mauve hairs but can be ± glabrous; rachis coloured like stipe; pinnae opposite or subopposite, triangular-lanceolate, 25–35(–45) cm long, 15–20(–25) cm wide, lower ones petiolate, upper ones sessile and adnate; ultimate lobes rounded. Sori marginal, ± 2 mm long, linear or curved, borne in and around sinuses of ultimate segments; indusium pale, glabrous. Fig. 7.

UGANDA. Mengo District: Kyagwe, 1 km N of Maigwe, 11 Sept. 1969, *Faden, Evans & Lye* 69/1046!

TANZANIA. Lushoto District: E Usambaras, Gonja, Sept. 1893, *Holst* 4233!; Kilosa District: Ukaguru Mts, Mamiwa forest reserve, 2 km N of Mandege, 2 Aug. 1972, *Mabberley* 1365!; Iringa District: Mwanihana forest reserve, above Sanje village, 8 Sept. 1984, *D.W. Thomas* 3683!

DISTR. U 4, T 3, 6, 7; Guinea south to Angola, Congo (Brazzaville), Central African Republic, Congo (Kinshasa), Zambia and Madagascar

HAB. Stream sides in evergreen forests, *Cyathea* forest, swampy places, granite rocks in stream valleys, roadside banks; 950–1600 m

SYN. *Antiosorus occidentalis* (Baker) Kuhn, Festschr. 50 Jahr. Jubil. K. Realschule Berlin (Chaetopt.): 347 (1882), *nom invalid.*

Anisosorus occidentalis (Baker) C. Chr. in Perrier, Cat. Pl. Madag., Pterid.: 54 (1932) & in Dansk Bot. Arkiv 7: 138, t. 54/10–11 (1932); Tardieu in Mém. I.F.A.N. 28: 85 (1953); Alston, Ferns W.T.A.: 33 (1959); Tardieu, Fl. Cameroun 3, Ptérid.: 104 (1964) & Fl. Gabon, Ptérid.: 78 (1964)

7. ODONTOSORIA

(C. Presl) Fée, Mém. Foug. 5: 325 (1852); Maxon in Contr. U.S. Nat. Herb. 17: 157–168, tt. 2–5 (1913); Kramer in Kubitzki et al., Fam. Gen. Vasc. Pl. 1: 91 (1990), pro parte excl. *Sphenomeris* Maxon

Davallia J. Sm. sect. *Odontosoria* C. Presl, Tent. Pterid.: 129 (1836)

Terrestrial ferns with scandent fronds; rhizomes mostly wide-creeping, densely clothed with slender hair-like scales. Fronds of indeterminate growth, 2–4-pinnate; pinnae opposite or alternate, the ultimate pinnules small, linear to cuneate-flabellate and variously lobed; stipe and rachis often somewhat woody, usually spiny in the

---

* Schelpe (1977) gave the K sheet of *Welwitsch* 132 as holotype and this has been accepted as a lectotypification; Baker gave other localities – Mt Cameroun and Fernando Po – in his original description.

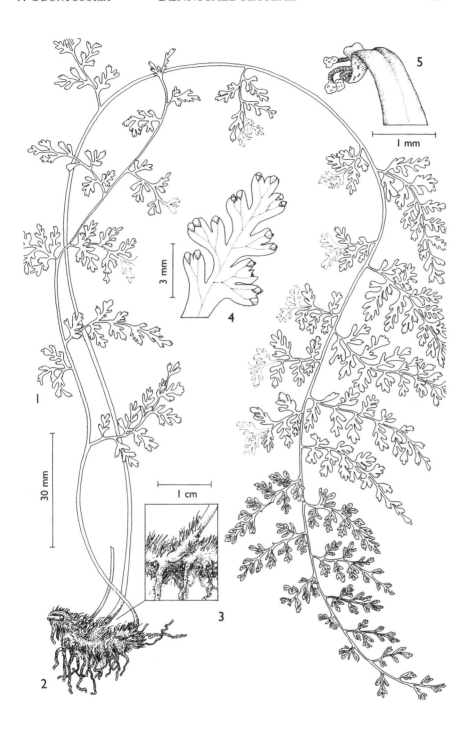

FIG. 8. *ODONTOSORIA AFRICANA* — **1**, habit; **2**, rhizome; **3**, rhizome showing scales; **4**, tip of fertile pinnule; **5**, tip of pinnule with indusium, after release of spores. 1, from *Rose* 10284; 2, 3, 5 from *Loveridge* 444; 4 from *Thomas* 2409. Drawn by Kathleen McKeehen.

American species. Sori terminal on single veins, immersed in tissue; indusium joined or partially joined to the opposed frond lobule to form an urceolate involucre, opening outwards; spores trilete.

About a dozen species, 2 occurring in the Old World, one in Africa, the other in Madagascar. Kramer (1990) combined *Sphenomeris* with this genus but it is kept separate here.

O. africana *Ballard* in K.B. 12: 143, fig. 1 (1957); Kramer in F.C.B., Lindsaeaceae: 4, photo 1, map 1 (1971) & in B.J.B.B. 42: 312, t. 8 (1972); Pic. Serm. in B.J.B.B. 53: 270 (1983). Type: Uganda, Kigezi District, Mushungero, *A.S. Thomas* 2409 (K!, holo. & iso., KAW, P!, iso.)

Climbing fern to 1.5 m, without spines; rhizome shortly creeping, scales dark brown, rigid, 4–6 cells wide at base, abruptly narrowing into a long jointed hair at apex. Stipes reddish or straw-coloured, 13.5–45(–65) cm long, channelled above; rachis similar, flexuous; lamina 3–4(–5 fide Ballard)-pinnate and pinnatifid, lanceolate-ovate or narrowly oblong in outline, 0.5–1.5(–4) m long, 18–38 cm wide; pinnae alternate, spreading, lanceolate to ovate-lanceolate or triangular; primary pinnules triangular to lanceolate; secondary and tertiary pinnules variable in outline but ultimate pinnule forked or flabellately or pinnately deeply cut; ultimate segments narrowly oblong, 1–3 mm long, 0.5–0.75 mm wide, mostly 1-veined; sorus elliptic to obovate, about half as wide as segment; indusial flap about as long as lamina, edge of lamina and indusium entire. Spores trilete, globose-tetrahedral. Fig. 8.

UGANDA. Kigezi District: Bukimbiri, Oct. 1947, *Purseglove* 2492! & Luhizha to Itigezi, Impenetrable Forest, 30 Sept. 1961, *Rose* 10318! & L. Mutanda, above Mushongero, 31 Jan. 1939, *Loveridge* 444!
DISTR. U 2; Congo (Kinshasa) (Kivu), Rwanda, Burundi
HAB. On rocks, in scrub, on red mud banks; 1800–2700 m

SYN. [*O. melleri* sensu Brause & Hieron. in Z.A.E. 2: 6 (1910), *non* (Hook.) C. Chr.]
   [*Stenoloma chinense* (L.) Bedd. var. *divaricatum* sensu Demaret in Expl. Hydrobiol. Lac Tanganyika 4 (2): 33, fig. 2 (1955), *non* (H. Christ) Alston]

8. SPHENOMERIS

Maxon in J. Wash. Acad. Sci. 3: 144 (1913) & in Contr. U.S. Nat. Herb. 17: 159 (1913); J.K. Morton in Taxon 8: 29 (1959), *nom. conserv.*

Erect or ascending ferns of determinate growth with short to very long slender rhizomes; scales very narrow, often hair-like and one cell thick. Fronds subfasciculate; stipes not jointed; lamina 3–4-pinnate or pinnatifid with alternate pinnae, the ultimate segments strongly cuneate, bifurcate; veins free. Sori single or 2-4 together, terminal at or near the truncate apices of the segments; indusium similar in texture to opposed segment-margin, flattish, pocket-like, attached at base and sides, single at clavate apices of the veins or, if joined, borne upon a translucent receptacle connecting these.

3 species in New World tropics and 8 in the Old World including 2 in Africa (the rest in Asia to Polynesia).

S. afra *Kramer* in B.J.B.B. 41: 353, fig. 1 (1971) & in F.C.B., Lindsaeaceae: 3, fig. 1, map 1 (1971) & in B.J.B.B. 42: 310 (1972); Schelpe in C.F.A., Pterid.: 112 (1977). Type: Angola, Xá-Sangue, *Young* 1118 (BM!, holo., U, iso.)

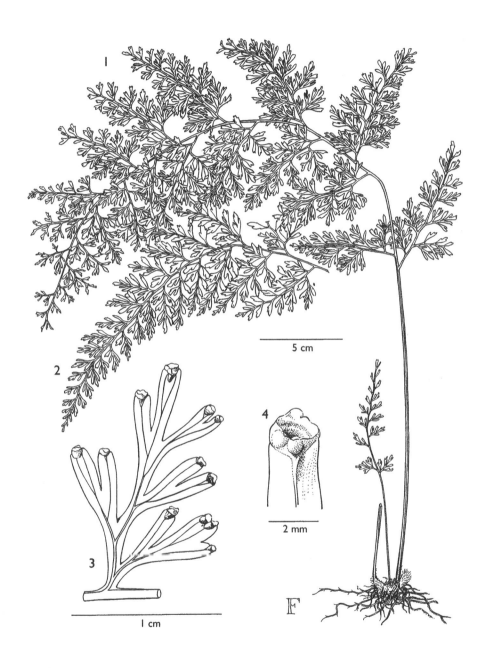

FIG. 9. *SPHENOMERIS AFRA* — **1**, habit, × ¹/₂; **2**, part of frond, × ¹/₂; **3**, pinnule, × 4; **4**, tip of
fertile pinnule, × 7. 1–2 from *Thollon* 5 Nov. 1909, Gabon material; 3–4 from *Dawe* 1921,
Angola material. Drawn by Monika Shaffer-Fehre.

Rhizome shortly creeping, 2–3 mm in diameter with very narrow chestnut brown scales only 1–2 cells wide, save at extreme base where 3-celled. Fronds close or tufted; stipe straw-coloured, 10–40 cm long; lamina thin, narrowly oblong, 0.3–1.5 m long, 8–50 cm wide, 2–4-pinnate and pinnatifid; primary pinnae in about 10–20 pairs, the longest up to 40 cm long, 12 cm wide, with petiolules up to 2 cm long; ultimate pinnules almost linear, up to 4 mm long, 0.75–1.25 mm wide, slightly divaricate, not dilated at the apex, uninerved, each with only 1 sorus; rachis straw-coloured. Indusium pocket-like, attached by the side, 0.3–0.5 mm long and wide. Spores trilete. Fig. 9.

TANZANIA. Rungwe District: Kiwira [Kibila], 7 Oct. 1911, *Stolz* 929!
DISTR. **T** 7; São Tomé, Congo (Brazzaville), Cabinda, Congo (Kinshasa), Angola
HAB. River valley; 1350 m

SYN. [*Odontosoria chinensis* sensu Bonap., Not. Ptérid. 14: 269 (1923), *non* (L.) Sm.]
    [*Stenoloma chinense* (L.) Bedd. var. *divaricatum* sensu Tardieu in Mém. I.F.A.N. 28: 65, t. 9/1–2 (1953)]

NOTE. Kramer has identified some São Tomé material as a probably undescribed species; he also makes the perennial error of referring Stolz's specimen to Malawi where he never collected. The other São Tomé taxon is *O. chinensis* var. *divaricatum* H. Christ in Chevalier in Journ. de Bot., sér. 2, 2: 23 (1909). Type: São Tomé, towards the Pic, after Lagua Amelia, *Chevalier* 14309 (P, syn.).

9. LINDSAEA

Sm. in Mém. Acad. Roy. Sc. Turin 5: 401, fig. 4 (1793); Bernh. in Neues J. Bot. (Schrad.)1(2): 34 (1806); Kramer in Kubitzki et al., Fam. Gen. Vasc. Pl. 1: 92 (1990)

Terrestrial or epiphytic with short to long creeping rhizomes bearing ovate to narrowly lanceolate or less often hair-like scales. Fronds with stipes very short to long; lamina thin to coriaceous, usually glabrous, very diverse, simple, pinnatifid, or once- to several-pinnate, the ultimate divisions often dimidiate, entire or lobed, wedge-shaped to linear or lanceolate, oblong, reniform etc.; veins issuing from posterior margin in dimidiate pinnules or simple to 1–several times forked, ending behind the margin, free or reticulate without free included veinlets. Sori terminal, continuous or interrupted, linear or oblong, on single veins or on a commissure joining all vein ends of a segment; indusium short to elongate, opening outwards; paraphyses small, few-celled or wanting.

About 150 species in the tropics, and in some temperate areas mainly in the southern hemisphere and Japan; only 3 appear to have been found in Africa. The species are remarkably diverse but attempts to split the genus have not been accepted by Kramer. The genus is sometimes placed in the Lindsaeaceae (or Lindsayaceae of Alston).

Fronds 2– ± 3-pinnate; ultimate divisions 4–13 × 3–7 mm · ·   1. *L. madagascariensis*
Fronds simply pinnate (or rarely simple); ultimate divisions
    (pinnae) 10–22 × 0.4–2.5 cm · · · · · · · · · · · · · · · · ·   2. *L. ensifolia*

1. **L. madagascariensis** *Baker* in J.L.S. 16: 198 (1877) & in Hook., Ic. Pl. 17, t. 1629 (1886); C. Chr. in Dansk Bot. Arkiv 7: 81, t. 27/9–11, as '*Lindsaya madagascariensis*' (1932); Kramer in B.J.B.B. 42: 318 (1972). Type: Madagascar, 'interior' (probably near Antananarivo), *Gilpin* (K!, holo.)

Rhizome sparsely branched, 1–1.5 mm in diameter; scales pale brown, needle-like, up to 1 mm long, uniseriate or biseriate at extreme base, soon ± wearing off. Fronds with stipes 10–20 cm long; lamina triangular in outline, 7–15 cm long, 5–15 cm wide, acute to sharply acuminate, usually 2–3-pinnate; pinnae in 2–6 pairs, lanceolate, 4–9 cm long, 1–3 cm wide, subobtuse or slightly acuminate with stalks 1–3 mm long;

FIG. 10. *LINDSAEA ENSIFOLIA* — **1**, habit, × ¹/₂; **2**, enlargement of fertile frond, × 3. Both from *Buchanan* 160. Artist not traced. From F.Z.

secondary pinnae in 5–7 pairs; ultimate pinnules very variable according to degree of dissection of the lamina. Sori on 1–4 vein ends, interrupted by marginal incisions; indusium subentire to erose, 0.3–0.6 mm wide, not reaching margin.

var. **madagascariensis**

Fronds not very finely divided, the ultimate segments not linear but cuneate, 4–13 mm long, 3–7 mm wide.

TANZANIA. Njombe District: Lupembe, Ditima Mt, 2 Apr. 1931, *Schlieben* 572 (photo!)
DISTR. **T** 7; otherwise known only from Madagascar
HAB. Primary forest; over 1500 m

SYN. *Schizoloma madagascariense* (Baker) Kuhn, Chaetopt.: 346 (1882)
    Sphenomeris madagascariensis (Baker) Tardieu in Amer. Fern Journ. 48: 34 (1958) & in Fl.
      Madag. 5 (1): 24 (1958)

NOTE. Var. *davallioides* Baker from Madagascar has very finely divided fronds (see C. Chr. in Dansk Bot. Arkiv 7: 81, t. 27/12).

2. **L. ensifolia** *Sw.*, J. Bot. (Schrad.) 1800 (2): 77 (1801); Kuhn, Fil. Afr.: 67 (1868); Baker, Fl. Maurit. & Seych.: 472 (1877); Schelpe in Journ. S. Afr. Bot. 35: 134 (1969) & F.Z., Pterid.: 139, t. 43 (1970); Kramer in Fl. Males. ser. II, 1: 211 (1971) & in B.J.B.B. 42: 328 (1972); Schelpe & Diniz, Fl. Moçamb., Pterid.: 141(1979); W. Jacobsen, Ferns S. Afr.: 294, figs. 210/a–b (1983); Schelpe & N.C. Anthony, F.S.A., Pterid.: 151, fig. 48/1, map 127 (1986); J.E. Burrows, S. Afr. Ferns: 182, t. 29/4, fig. 38/182, map (1990). Type: Mauritius, collectors not known (S-PA, holo.)

Rhizome creeping underground (*fide* Wingfield), 1–2.5(–3) mm diameter; scales pale to dark reddish brown, lanceolate to triangular, up to 2.5 mm long and to 5-seriate at the base. Fronds spaced, 70–90 cm tall in East Africa; stipes brown and shiny, 10–36 cm long; lamina very variable, if simple then linear or lanceolate up to 10 cm long and 0.3–1 cm or 1.5–3 cm wide, if pinnate 15–48 cm long and up to 22 cm wide; pinnae in 2–8(–12) pairs, linear to lanceolate or narrowly oblong, 10–22 cm long, 0.4–2.5 cm wide, broadly to narrowly cuneate at base and ± symmetrical. Sori continuous, with pale brown linear indusium 0.3–0.5 mm wide.

subsp. **ensifolia**; Kramer in Fl. Males. ser. II, 1: 212 (1971)

Upper pinnae not much reduced, the lamina with a free terminal similar pinna; rachis sharply bi-angular, 0.4–0.5 mm wide, entire; pinnae of sterile fronds subentire to mostly serrate at margin. Indusium linear. Fig. 10.

TANZANIA. Uzaramo District: 1.5 km SW of Kisarawe, near Kimani stream below road to Maneromango, 26 Dec. 1976, *Wingfield* 3739!; Pemba: Makongeni [Makondeni], 22 May 1934, *Vaughan* 2201!
DISTR. **T** 6; **P**; Guinea, Ghana, Nigeria, Gabon, Mozambique, South Africa, Madagascar, Seychelles, Mauritius, Réunion, India, Sri Lanka, Malesia to W Melanesia etc.
HAB. Open seepage area on kaolin in forest, 'water fern' *fide* Vaughan; 0–150 m

SYN. *Schizoloma ensifolium* (Sw.) J. Sm. in Journ. Bot. [Hook.] 3: 414 (1841); V.E. 2: 21, fig. 19 (1908); C. Chr. in Trans. Linn. Soc. Ser. 2, 7: 415 (1912); Sim, Ferns S. Afr., ed. 2: 130, t. 39 (1915); C. Chr. in Dansk Bot. Arkiv 7: 79 (1932); Alston & Schelpe in Journ. S. Afr. Bot. 18: 157 (1952); Tardieu in Mém. I.F.A.N. 28: 64 (1953)
    Schizolegnia ensifolia (Sw.) Alston in Bol. Soc. Brot. sér. 2, 30: 24 (1956); Tardieu, Fl. Madag. 5 (1): 31 (1958); Alston, Ferns W.T.A.: 44 (1959); Tardieu, Fl. Gabon 8, Ptérid.: 81, t. 14/1–2 (1954)

NOTE. The variation is extensive but with numerous interconnecting forms; Kramer gives the very lengthy synonymy (1971, 1972). Two other subspecies occur from Malesia to the Pacific.

INDEX TO DENNSTAEDTIACEAE

New names validated in this part

GEOGRAPHICAL DIVISIONS OF THE FLORA

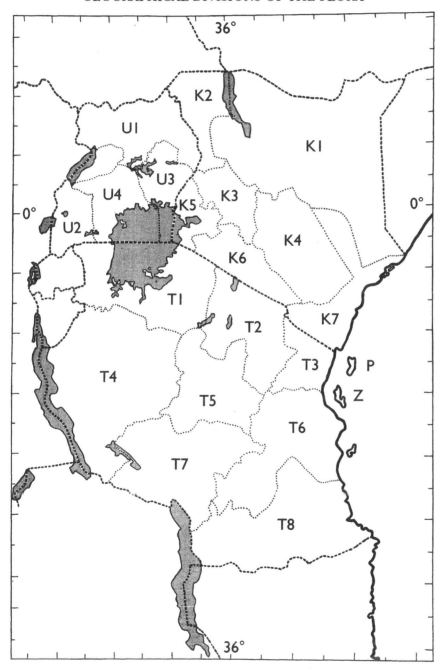